ENVIRONMENTALLY AND SOCIALLY
SUSTAINABLE DEVELOPMENT
Rural Development

Intellectual Property Rights in Agriculture

The World Bank's Role in Assisting Borrower and Member Countries

Uma Lele
William Lesser
Gesa Horstkotte-Wesseler
Editors

The World Bank
Washington, D.C.

Cover photographs. Top: Potato harvest in the Andes, International Potato Center (CIP), Peru. Bottom: Biotechnology lab, International Maize and Wheat Improvement Center (CIMMYT), Mexico. Photographs by permission of the Consultative Group on International Agricultural Research (CGIAR) Secretariat.

Uma Lele is an adviser in the Rural Development Department of the World Bank. William Lesser is a professor in the Agricultural Resource and Managerial Economics Department, Cornell University. Gesa Horstkotte-Wesseler is a junior professional officer in the Rural Development Department of the World Bank.

Library of Congress Cataloging-in-Publication Data

Intellectual property rights in agriculture : the World Bank's
    role in assisting borrower and member countries /
    Uma Lele, William Lesser, and Gesa Horstkotte-Wesseler, editors.
        p.   cm. — (Environmentally and socially sustainable
    development. Rural Development)
        ISBN 0-8213-4496-X
        1. Intellectual property—Developing countries.   2. Agriculture—
    Research—Law and legislation—Developing countries.   3. World
    Bank.   I. Lele, Uma J.   II. Lesser, William.   III. Horstkotte-
    Wesseler, Gesa, 1965–   .   IV. Series: Environmentally and socially
    sustainable development series. Rural development.
    K1401.I567   1999
    630'.7'2—dc21   99-15437
                                                                        CIP

Contents

Boxes

Tables

Foreword

The growing involvement of the private sector in worldwide agricultural biotechnology research and the WTO requirements for member countries to establish intellectual property rights (IPR) regimes by 2000 (some by 2005) make the issue of IPR an urgent priority in and for developing countries.

As the largest lender for agricultural research the World Bank has an important role to play in assisting its client countries to strengthen their IPR regulations and enforcement policies. The challenge of assisting developing countries is made quite complex by the rapid changes in trade practices, the wide variation in the stage of development of the Bank's client countries, and the evolution of global agreements on the rights of different stakeholders with regard to plant genetic material. Private sector investment in agricultural research crucially depends on the protection of intellectual property.

The World Bank decided to organize a major workshop on IPRs, bringing together a number of informed individuals representing major institutions with a stake in IPRs. The Bank took this initiative because there is increasing concern about the rapidly growing role of the private sec-

tor in agricultural research among the Bank's client countries and because there has been a steady drop in recent years in the rate of growth of public sector funding in agricultural research. This is happening despite the potential threat to nearly 100 million resource-poor rural households dependent on farming, who are served by the public sector. Public/private partnerships in research in the future would need to serve the interest of these farmers.

It was felt vital to address the challenges surrounding IPRs to protect food security and to ensure uninterrupted supply of new technologies to resource-poor farmers.

This volume contains valuable insights into the complex area of intellectual property rights in agriculture and will help give urgently needed direction to the Bank and to its member and borrower countries.

Ian Johnson
Vice President
Environmentally and Socially
Sustainable Development

Preface

Intellectual Property Rights (IPR) in agriculture and the possible role that the World Bank might play in assisting its developing country borrowers/members were the main topics of a workshop held June 11–12, 1998, in Washington, D.C.

The goal was to develop a better understanding of the perspectives and the challenging problems some of the key stakeholders face with regard to agriculturally related IPRs and to define possible roles for the Bank to help its borrower/member countries. The emphasis was on developing countries, and areas of particular interest included problems faced by these countries in implementing effective IPR systems and in identifying areas where further research and operational support by the Bank, through research, policy advice, and lending activities, would contribute to this goal.

The introductory chapter discusses gaps in knowledge about the effects of IPR and priority research areas, approaches to strengthen national IPR systems in the short and long terms, identification of practical hurdles to implementing effective IPR systems, and effective incorporation of IPR within the scientific community, including the Consultative Group on International Agricultural Research and the national agricultural research systems of developing countries.

Recommendations for possible research and operational support by the World Bank are discussed in the various chapters, and summarized in the concluding chapter.

Uma Lele, William Lesser,
and Gesa Horstkotte-Wesseler

Acknowledgments

This publication grew out of a workshop on intellectual property rights (IPR) and the role of the World Bank. The idea for the workshop and this publication originated in discussions among Alberto D. Portugal, president of EMBRAPA and the head of the Brazilian National Agricultural Research System; Francisco J. B. Reifschneider, director of EMBRAPA's Secretariat for International Cooperation; and Uma Lele of the World Bank, who manages the Brazil Agricultural Technology Development Project (PRODETAB) funded by the Bank. These discussions helped clarify Brazil's needs in agriculturally related IPR, the likely effect of IPR on private sector investments, and public/private partnerships in agricultural research in Brazil. The discussions also focused on how the World Bank might assist Brazil to establish a sound IPR regime.

Ismail Serageldin, chairman of the Consultative Group on International Agricultural Research (CGIAR) and then–vice president for Environmentally and Socially Sustainable Development at the World Bank, recognized immediately the importance of the rapidly emerging challenge of IPRs in the World Bank's lending function, as distinct from the Bank's role in the CGIAR. He lent strong support to organizing a workshop that would explore implications for the Bank as a lender and policy advisor to its member and client countries. Carlos Primo Braga, a leader in the area of IPR in the World Bank, encouraged exploration of the special case of agriculture. Gregory Ingram, director of the World Bank's Research Committee, and the Agricultural Knowledge Information Systems (AKIS) thematic group in the World Bank's Rural Development Family (RDV) provided funds for the workshop and for this publication. Alex McCalla, director of RDV, supported publication of this volume, and RDV partially contributed to its cost.

Robert Sherwood offered extensive comments on an early draft of the introductory chapter, and John Barton helped with some of the implications of the workshop for the World Bank. Reginald MacIntyre provided valuable editorial input, as did Virginia Hitchcock and Alicia Hetzner. Jason Yauney provided strong staff support throughout the project. This volume was desktopped by Gaudencio Dizon.

Contributors

Derek Byerlee, Rural Development Network, World Bank, 1818 H Street N.W., Washington, D.C. 20433 U.S.A.

Joel I. Cohen, Intermediary Biotechnology Service, International Service for National Agricultural Research, P.O. Box 93375, 2509AJ The Hague, The Netherlands.

Frederic H. Erbisch, Office of Intellectual Property, Michigan State University, East Lansing, MI 48824 U.S.A.

Cesar Falconi, Intermediary Biotechnology Service, International Service for National Agricultural Research, P.O. Box 93375, 2509 AJ The Hague, The Netherlands.

Gesa Horstkotte-Wesseler, Rural Development Family, World Bank, 1818 H Street N.W., Washington, D.C. 20433 U.S.A.

S. M. Ilyas, Indian Council of Agricultural Research (ICAR), Krishi Bhavan, New Delhi 110001, India.

John Komen, Intermediary Biotechnology Service, International Service for National Agricultural Research, P.O. Box 93375, 2509AJ The Hague, The Netherlands.

Uma Lele, Rural Development Family, World Bank, 1818 H Street N.W., Washington, D.C. 20433 U.S.A.

William Lesser, Department of Agricultural, Resource, and Managerial Economics, Cornell University, Ithaca, New York, U.S.A.

Peter Ninnes, Centro Internacional de Mejoramiento de Maiz y Trigo, Apartado Postal 6-641, Lisboa 27, 06600 Mexico DF, Mexico.

Gabrielle J. Persley, Bioalliance Australia Limited, P.O. Box 1101, Toowong, Brisbane, Queensland, Australia.

Timothy G. Reeves, Centro Internacional de Mejoramiento de Maiz y Trigo, Apartado Postal 6-641, Lisboa 27, 06600 Mexico DF, Mexico.

David L. Richer, Hoechst Schering AgrEvo GmbH, Hoechst Works, K801, D-65926 Frankfurt am Main, Germany.

Maria José Amstalden Sampaio, Brazilian Agricultural Research Corporation (EMBRAPA), Caixa Postal 040315, 70770-901 Brasilia DF, Brazil.

Richard H. Shear, Monsanto Company, 800 North Lindbergh Blvd., St. Louis, MO 63167 U.S.A.

Elke Simon, Hoechst Schering AgrEvo GmbH, Hoechst Works, K801, D-65926 Frankfurt am Main, Germany.

Jayashree Watal, Institute for International Economics, 11 Dupont Circle, N.W., Washington, D.C. 20036-1207 U.S.A.

Abbreviations and Acronyms

ARIPO — African Regional Intellectual Property Organization (Harare)

CBD — Convention on Biological Diversity

CGIAR — Consultative Group on International Agricultural Research

CGRFA — Commission on Genetic Resources for Food and Agriculture (FAO)

CIMMYT — Centro Internacional de Mejoramiento de Maiz y Trigo (International Center for Maize and Wheat Improvement)

CIP — Centro Internacional de la Papa (International Potato Center)

DNA — Deoxyribonucleic acid

ELISA — Enzyme-linked immunosorbent assay

EMBRAPA — Empresa Brasiliera de Pesquisas Agropecuarias (Brazilian Agricultural Research Corporation)

EPO — European Patent Office

ESDAR — Environmentally Sustainable Development Agricultural Research and Extension Group (World Bank)

FAO — Food and Agriculture Organization of the United Nations

GATT — General Agreement on Tariffs and Trade (now WTO)

GDP — Gross domestic product

IARC — International agricultural research center

ICRIER — Indian Council for Research on International Economic Relations

ICRISAT — International Crops Research Institute for the Semi-arid Tropics

IITA — International Institute of Tropical Agriculture

INIBAP — International Network for the Improvement of Banana and Plantain

INPADOC — International Patent Documentation Center

INPI — National Institute for Industrial Protection (Brazil)

IPGRI — International Plant Genetic Resources Institute

IPM — Integrated pest management

IPR — Intellectual property rights

IRRI — International Rice Research Institute

ISNAR — International Service for National Agricultural Research

KARI — Kenya Agricultural Research Institute

LGC — Land grant college

MTA — Material transfer agreement

NARO — National agricultural research organization

NARS — National agricultural research system

NGO	Nongovernmental organization	RFLP	Restriction fragment length polymorphism
OAPI	Organisation Africaine de la Propriete Intellectuelle (Yaounde)	TRIPs	Trade-related aspects of intellectual property rights
OECD	Organisation for Economic Co-operation and Development	UNCTAD	United Nations Conference on Trade and Development
ORSTOM	Institut Francais de Recherche Scientifique pour le Developpement en Cooperation (France)	UPOV	Union Internationale pour la Protection des Obtentions Vegetales (Convention for the Protection of New Varieties of Plants)
PBR	Plant breeders' rights		
PCT	Patent Cooperation Treaty		
PGRFA	Plant genetic resources for food and agriculture	USDA	United States Department of Agriculture
R&D	Research and development	WIPO	World Intellectual Property Organization
RAFI	Rural Advancement Foundation International	WTO	World Trade Organization

1. Intellectual Property Rights, Agriculture, and the World Bank

William Lesser, Gesa Horstkotte-Wesseler, Uma Lele, and Derek Byerlee

After initial reluctance, many developing countries are modifying their intellectual property systems. Those that are members of the World Trade Organization (WTO) must make intellectual property rights (IPR) regulations conform with the TRIPs (Trade-Related Aspects of Intellectual Property Rights) agreement by January 1, 2000. Least-developed countries have an additional five years to meet the same goal. All countries have until 2005 to implement protection for new areas of technology. More generally, globalization, expansion of international trade and competition, increased foreign direct investment, and growing international pressure, particularly from the United States, are compelling many developing countries to enhance their IPR protection. At the same time these countries face a rapid rise in the number of IPR-related disputes and sanctions involving their own industries, research scientists, plant genetic material, and uses of indigenous knowledge (Lourie 1998). One school of thought maintains that developing countries with effective IPRs will attract more research and development (R&D) spending, particularly from the private sector. A second widely held view disputes this conclusion, maintaining that, at the extreme, IPR amounts to economic colonialism.

Legislation without implementation is of little value. Capacity to implement IPR legislation is essential and urgently needed. Implementation requires a complex and sophisticated infrastructure that is lacking in most developing countries. Even with the best of intentions, and whatever level of protection is sought, it will take time for developing countries to establish effectively functional protection (that is, reaching a point where the private sector develops sufficient confidence to increase investments in science and technology *in developing countries*, as well as to transfer key technologies there).

Complicating this process, and any consideration of a supportive role for the World Bank, is the existence of large gaps in our knowledge of IPRs in agricultural technology. The issues of IPRs in agriculture are complex, more than in any other sector, particularly pertaining to developing countries. They involve a range of stakeholders with a diverse set of views, bargaining positions, and vulnerabilities, and are therefore laden with controversy and emotion.

Uniqueness of Agriculturally Related IPRs

Agriculture is unique because of its diversity and location-specific requirements, necessitating adaptation of technologies to a range of agroecological conditions. A large number of poor households in developing countries derive their livelihood from resource-poor areas with difficult agroclimatic conditions. Ensuring their access to technologies is therefore crucial for poverty alleviation. A combination of the rapid advances in science and the number of applications of proprietary rights has led to an unprecedented growth in private sector investment in agricultural technology, estimated at US$7 billion in 1996 (ISAAA 1996). This has resulted, in a matter of just a few years, in unforeseen increases in acreage under disease- and pest-resistant transgenic materials, mostly in industrial countries. In the United States in 1998, nearly 50 percent of cotton, 40 percent of soybean, and 20 percent of corn acreage were planted to trans-genic varieties (Kilman and Warren 1998). Some researchers argue that this is in part a result of the public sector in industrial countries placing a lower priority on agricultural research during a period of constrained budgets (Woteki 1998).

The green revolution technology of the 1960s of high-yielding cereal seed varieties resulted in a shift in productivity growth and commercialization of agriculture in many developing countries, an experience that could suggest what will happen with biotechnology. There are, however, several key differences that could change the outcome. Public goods research, at national and international levels, played a crucial role in providing free and open access to green revolution technology for a large number of subsistence farmers. The scientific work of the Consultative Group on International Agricultural Research (CGIAR) was anchored strongly in the *public sector research* systems of industrial countries, with much of the early improved germplasm evolving from research initiated by the public sector in industrial countries. Presently, public sector agricultural research is more limited at all levels.

Despite the impressive achievements of the green revolution, nearly 1 billion poor people in developing countries still derive their sustenance from agriculture using their own traditional plant genetic material. It is an important mission, and a challenge for public sector researchers of developing countries, the CGIAR system, and the World Bank to ensure that these households have free access to these improved technologies. Developing technologies for these farmers requires considerable capital, including the collection, characterization, and enhancement of genetic material. The benefits take time to materialize and are difficult to capture, making such investments unattractive to the private sector. Rapid population and income growth in developing countries has resulted in a higher demand for food and fiber. This can be met only by continued increases in agricultural productivity, under conditions of increasingly scarce and often deteriorating land and water resources. The agricultural sector of developing countries thus faces a dual challenge. The first challenge is to sustain

and enhance the quality of natural resources, including germplasm, as the only way to meet the growing domestic demand and international competition. The second challenge is to ensure continued free access for the poor to emerging technologies while conserving vital germplasm for use by plant breeders. In both instances, the new bioscience would be of considerable value if access to the poor were provided.

Many large developing countries are increasing their investments in agricultural research while introducing major reforms in their national research systems. The reforms emphasize a shift to increased partnerships between the public and private sectors at the national and international levels. The World Bank presently supports these reforms through loans and credits in Brazil, India, and Indonesia. Similar reforms are possible with World Bank assistance in China.

Brazil, India, and Indonesia are also experiencing large private investments in their seed industries. The acquisition of some of the largest seed companies (for example, Agroceres in Brazil) by multinational corporations (for example, Monsanto) may well have a more profound effect on public sector research strategies of the national agricultural research systems (NARS) by determining the long-term comparative advantage of public vis à vis the private sector than the current reforms of the public research system per se.

Public sector support by industrial countries to the international and national agricultural research systems of developing countries is in a steady decline, while the challenges facing international agriculture have become more diverse. These range from the maintenance of biodiversity and the management of natural resources to poverty alleviation and productivity enhancement (Lele 1998). *Expanding proprietary science, market segmentation by product, geographical areas and size class of farmers, between industrial and developing countries, and within developing countries, is increasingly the name of the game in R&D. An important goal that must be pursued is access to genetic resources for the global common good, despite restrictions imposed by national and international public research systems as they seek enhanced protection through IPRs for themselves.*

Some public sector researchers, because of limited resources, have sought support from private firms, thereby blurring the role of the public sector. A recent survey indicated that (a) 12 percent of U.S. university scientists conducting private sector-funded research reported generating trade secrets; (b) 24 percent indicated that research could not be published without consent of the sponsor; (c) 44 percent reported that to some degree industry support undermines intellectual exchange; (d) 30 percent stated that future commercial application had influenced their choice of research topics; and (e) 70 percent felt that such support shifts emphasis too much towards applied research (*The Lancet* 1993).

These issues are of relevance to developing countries and the CGIAR centers because partnerships are taking many new forms, including:

- Private investment in input industries in developing countries (for example, Monsanto's purchase of Agroceres)
- Private firm support of research of the type previously financed by the public sector in industrial and developing countries (for example, Novartis' rice genome project; see below)
- Developing country collaborative research in industrial countries (for example, Brazil's emerging LABEX program with the U.S. Department of Agriculture (USDA) and U.S. universities supported by the World Bank and possibly in the future with France and Japan)
- CGIAR collaboration with industrial countries' research institutions and universities, the multinational private sector, and developing countries.

These new partnerships have raised issues for the public and private sectors: on the one hand, maintaining control of developments and recouping investments, and on the other, fulfilling the special public sector mission.

Public sector researchers in industrial and developing countries increasingly find their work controlled by overlapping material transfer agreements (MTAs) and IPRs. At the level of products and processes IPRs are the major device employed. At the research level contracts known as MTAs are often used. A research MTA provides permission for research with a specified component technology, but permission for commercialization, if needed, will be based on subsequent negotiations.

As a case in point: a Cornell University scientist developed genetically engineered resistance to papaya ring spot virus, a devastating global disease for which there is no known natural resistance. A number of national programs, including those of Brazil and Thailand, invested in the development of variants with resistance to localized strains of the virus. However, it was later recognized that the final product incorporated eleven protected products of which Cornell owned just one. Negotiations over the commercial use of the other ten have not all been completed following three years of effort (Walter Haeussler, Director, Cornell Research Foundation, May 1998, personal communication). Particularly problematic for many researchers has been receiving permission for the use of Monsanto's popular 35S promoter. However, there are no clear alternatives to the current use of research MTAs. Negotiating commercialization agreements for every component product prior to initiating research is impractical given the failure rate of projects, and recreating protected materials like the 35S promoter is costly and time consuming. This situation of unclear rights has raised serious questions about the roles and opportunities of public sector agricultural research, and the optimal forms of partnerships.

For many years patent litigation in agricultural biotechnology appeared unlikely given high litigation costs and modest profitability, at least compared to the pharmaceutical industry (Eberhart and others 1998). Barton (1998), however, reports that this situation has changed radically with multiple disputes under litigation. Barton contends that the existence of a number of overlapping and competing broad patents contributes to oligopoly, the domination of the industry by a small number of firms (for example, Monsanto had an estimated worldwide agricultural biotechnology market share of 77 percent in 1998) (Reuters 1998). Oligopoly can lead to declining incentives for research investments by all firms, including industry leaders. The counter argument is that concentration offers sufficient profitability to permit continued research investments. Dominant firms are also in a position to purchase small start-ups for significant sums, encouraging new start-ups. This Schumpeterion theory, however, has not been empirically supported in other sectors, which show that the most innovative industries are characterized by moderate size and concentrations of firms (literature review in Scherer and Ross 1990). According to Barton (1998), an important policy issue for the United States and the world is to identify ways to *reverse* this oligopolistic trend and its consequences, while maintaining the strength of the IPR incentives to encourage research.

The search for a remedy to the oligopolistic trend was made more difficult because IPRs have had only an indirect effect on vertical control within the seeds sector. In the case of Monsanto, vertical control was extended largely through acquisitions—16 in 1997 and 2 major ones in 1998 prior to the Monsanto/American Home Products merger. The complete acquisition of Delta and Pine Lands (not approved at the time of writing), for example, will give Monsanto an estimated 80 percent share of the U.S. cotton seed market (Kilman and Warren 1998). Those acquisitions were driven in part, Lesser (1998b) argues, by the very high Monsanto share price and price-earning ratio, making stock-based acquisitions very economical at the time. Moreover, Monsanto's earlier success led to rapid run-ups in the price of seed company stocks—DeKalb, for example, jumped 10 times in market capitalization following Monsanto's initial purchase of 40 percent in February 1996 to full acquisition in 1998. IPRs contributed to the share value of Monsanto, but only indirectly as filtered through Monsanto's corporate strategy.

IPR Issues for CGIAR and NARS

The CGIAR is also facing issues of concentrated ownership, and having to negotiate commercial use arrangements for technologies licensed in as well as out. CGIAR leaders are also debating how best they can fulfill the mandate of continued free access of its technologies to developing countries in a situation of rapid growth in proprietary science. The genetic resources maintained in the gene banks of the Centers are held in trust for the world community in accordance with the Convention on Biological Diversity (CBD). Materials held in trust by the Centers are to be freely available in accordance with the CGIAR/FAO policy on plant genetic resources. Centers have not sought intellectual property protection on

improved materials and technologies, unless absolutely necessary to ensure access by developing countries. But in the face of rising IPR use, the Centers must be guided by the competing needs to (a) establish collaborative research with advanced laboratories; (b) ensure product development and distribution; and (c) forestall protection of CGIAR-generated technologies by others (Hawtin and Reeves 1998). IPRs have to be utilized without undoing the CGIAR's fundamental position regarding free access by developing countries to knowledge, technology, materials, and plant genetic resources. Presently, germplasm recipients of CGIAR center materials must, among other things, agree to require that all subsequent parties to whom the material is transferred also honor the same conditions. Alternative agreements are under discussion (for example, the use of umbrella agreements to cover multiple shipments to any one country). An important question is whether developing countries will be restricted in the use of CGIAR materials as those materials are more directly controlled under IPRs and related systems.

The CGIAR faces a number of issues that provide useful models for developing countries, including (a) the status of the *ex situ* materials acquired before the CBD was adopted, an issue to be resolved under the renegotiation of the International Undertaking (IU) on Plant Genetic Resources of the FAO Commission on Genetic Resources for Food and Agriculture (CGRFA); (b) the extent to which the designated material provided by the CGIAR has to be modified before it may be protected without violation of an MTA; (c) the extent to which patents can be applied to plants, varieties, or traits as distinct from "cells, organelles, genes or molecular constructs" to which the CGIAR guiding principles refer; (d) the controversial issue of "benefit sharing" (discussed below) of commercialized CGIAR materials with plant breeders on the one hand and the sources of the genetic material on the other; (e) regarding d above, the issue of incentives to invest in improvements to CGIAR materials in the absence of a clear guarantee of return, an issue raised particularly by the private sector seeking clarity in the CGIAR policy; (f) the cost of monitoring and enforcing unauthorized uses of CGIAR materials in a time of constrained support; and (g) ensuring that the CGIAR is not violating IPRs or signed MTAs in the cases of protected technology components.

IPRs and the TRIPs Agreement

In recent years the forms of IPR available, including those applicable to agriculture and aquaculture, have been enlarged to accommodate new fields of inventive activity. IPR is typically defined to include five major forms of protection: patents, plant breeders' rights (PBR), copyright, trademark, and trade secrets. More recent and specialized forms of protection include integrated circuits. For agricultural applications the principal forms of protection used include patents, PBR, and trade secrets. Extension of IPRs to new forms, such as genetically modified living organisms, has been accomplished, where permitted, by a reinterpretation of the scope of existing statutes, particularly patents.

The extent of IPR protection varies widely across countries. In the United States in particular a wide scope of protectable subject matter existed. Numerous other countries limited the scope of protectable inventions, as well as the length of protection and the conditions under which a license could be granted for using an invention without the developer's permission (compulsory licenses). As of 1988, about 50 countries excluded pharmaceuticals from patent protection, and 54 countries explicitly excluded plant varieties (WIPO 1990). Among the 37 members of the Convention for the Protection of New Varieties of Plants (UPOV) the international convention for PBR, only nine are developing countries (Argentina, Chile, Colombia, Ecuador, Mexico, Paraguay, South Africa, Trinidad and Tobago, and Uruguay). Other countries, including Brazil, Kenya, and Zimbabwe, have national laws and many are preparing to join UPOV in the near future.

Within the context of the Uruguay Round of GATT and the associated TRIPs Agreement, signatory countries are required to provide certain minimal levels of IPR protection (for details see Beier and Schricker 1996). For living forms applicable to agriculture, the major requirements include:
- Subject to the exclusions identified below, patents shall be available for any invention in all fields of technology (Section 5, Article 27(1))

- Contracting parties shall provide for the protection of plant varieties by patents and/or by an effective *sui generis* system (Section 5, Article 27(3b))
- Patents may, subject to a proviso, be prohibited in order to protect *ordre public* or morality, including to protect animal or plant life or health (Section 5, Article 27(2))
- Plants and animals other than microorganisms and "essentially biological processes for the production of plants and animals" may be excluded from patent protection (Section 5, Article 27(3b))
- Compulsory licenses may be issued, subject to limitations, in specific cases, including failure to make a licensing agreement after a reasonable offer, and subject to adequate remuneration, and to judicial review (Section 5, Articles 30 and 31)
- For process patents, the burden of proof of infringement may, in some specified circumstances, be shifted to the defendant to prove that the patented process was not used (Section 5, Article 34)
- Persons shall have the option of preventing others from using without permission information of commercial value so long as reasonable efforts have been made to keep it secret (Section 7, Article 39)
- Countries are to adopt extensive enforcement procedures which are "fair and equitable," are "reasoned" but "not unnecessarily complicated or costly" (Part III, Article 41).

In view of the January 1, 2000, deadline to implement these changes (plus five additional years allowed for the implementation of protection for product patents in new areas of technology), and with another five years for least-developed countries, a review of "implementation" is scheduled for 1999 (Article 71). Many among the most technologically advanced of the developing countries—including Brazil, Indonesia, and Mexico—have recently amended their patent laws and/or adopted PBR legislation. Others are progressing with the fulfillment of their commitments.

Evolving Developing Country View of IPRs

Developing countries fully realize the need to become more proactive in IPR. However, for many this is new and unknown territory. Initial reservations still remain among many stakeholders of research and technology in developing countries, including the NGO community. They feel that IPRs, based as they are on the Western concepts of property rights and jurisprudence, are either unethical or impractical in the cultural, historical, and institutional context of developing countries. They are costly to implement, requiring expensive IPR infrastructure and provision of protection, unfair because of the monopolistic tendencies of the private sector, and inequitable because of the existence of a large population of poor farmers who conserve the genetic resources upon which modern technology depends (Patel 1989; Shiva 1996). There have been concerns also that the IPR protection of plant varieties reduces genetic diversity (The Crucible Group 1994; Fowler 1998). A major issue in agriculturally related IPRs pertains to farmers' options to replant, save, exchange, and/or sell protected seed, as well as the potential availability of legal mechanisms to protect farmers' traditional varieties (Posey and Dutfield 1996). Some argue that even if IPR would be necessary at a more advanced stage of development in the future, they are not currently a priority (Konan and others 1995), and may be detrimental to the realization of certain objectives (for example, of imitating imported technologies to bridge the huge backlog of technologies generated abroad, and issues related to biosafety) (Doyle and Persley 1996). Some of these concerns will undoubtedly shape the ways in which developing countries fulfill their commitments to comply with the TRIPs agreement, particularly as it pertains to agricultural technology.

Concurrent to many of these developments has been the ratification of the CBD, which went into effect in December 1993. The CBD reaffirms the sovereign rights of States over their natural resources, in contrast to the earlier situation when plant germplasm was exchanged as a "Common Heritage of Mankind" among scientists in an open and generally unrestricted manner, with the exception of some plantation crops. However, the CBD does not itself provide a mechanism for controlling these resources, whereas patents may be used (and possibly misused) for developed products.

Developing countries have therefore become increasingly concerned that traditional indigenous knowledge of plant species is being appropriated by industrial countries through patents. This was demonstrated for turmeric, when a U.S. patent was granted for its wound-healing power, notwithstanding the fact that the powder of the turmeric plant is a "classic grandmother's remedy" in India which "has been applied to the scrapes and cuts of generations of children" (Agarwal and Narain 1996). The patent was subsequently rescinded.

Research Results Regarding the Roles and Functions of IPRs

IPRs may achieve a number of objectives. They can serve as

- A fundamental right of individuals to protect their intellectual property in much the same way as laws would protect other forms of property
- A way of ensuring incentives for innovation
- A way to induce the needed investments to develop and commercialize the invention
- An incentive to disclose information
- A mechanism for protecting the disclosure of partially-developed inventions, particularly during licensing talks
- An aid to technology transfer
- A way to enable the orderly development of broad prospects (Mazzoleni and Nelson 1997).

The effects of IPRs on areas such as economic growth, private R&D investment, technology transfer, and prices are poorly documented and hence controversial, particularly in agriculture. We reviewed the general literature on the economic and social effects of IPRs to identify key gaps in current knowledge in agriculture that require further research.

Economic Growth and Private Investment

Some observers of economic development as well as analysts of IPRs (Helpman 1993) consider leapfrogging through imitation as an essential aspect of the development process for latecomers. Others have argued that investment in innovation is arrested to a substantial extent by the absence of intellectual property protection, leading to slower growth (Gould and Gruben 1996). Even worse, due to the absence of a legal and institutional framework and a tradition of enforceable contracts related to intellectual property, many inventions made in developing countries are appropriated and commercialized by foreigners without adequate compensation for the inventor. For example, Sherwood, Scartezini, and Siemsen (1999) argue that Petrobras, Brazil's national oil company, made important inventions in deep-ocean platform drilling but failed to obtain patents, thereby making a gift of that technology to the oil companies of the world. Developing country inventors of sophisticated products routinely file for patents in industrial countries due to the high cost, long delays, and low value of the patents in their own countries.

The importance of IPR protection clearly varies with the level of technological activity. From two studies conducted for the World Bank, Mansfield (1995) concludes that "in relatively high-technology industries like chemicals, pharmaceuticals, machinery, and electrical equipment, a country's system of intellectual property protection often has a significant effect on the amount and kinds of technology transfer and direct investment." This argument would seem applicable to agricultural biotechnology. Braga (1995), in a literature review, concluded that enhanced IPRs in a country typically expanded imports of protected products and direct investment (see also Henderson, Voros, and Hirschberg 1996). These cross-sectional studies use aggregated trade and investment data, along with rather crude measures of IPR effectiveness. Specific effects of particular changes in IPR systems on product categories are not known. In particular little is known about the strategic decisions made by firms when introducing easily copied products into national markets.

The index of patent rights developed for 110 countries from 1960 to 1990 (Ginarte and Park 1997) suggests that high-income countries are on a higher level of the index than low-income countries. However, low-income African countries are the exception—their index is higher than that of middle-income countries, due to the adoption of the colonial system of IPRs. Evidence, however,

does not suggest high levels of R&D in African countries; the numbers of patents granted in all the southern African countries excluding South Africa in 1995 was only 727 (WIPO data are available on their web page, www.wipo.org). This suggests that other factors, including the availability of trained scientists and market size, may be the determining factors. Projecting from the past is hazardous, however, in a situation of dynamic science and a more liberal climate for markets in developing countries.

Notwithstanding the current emphasis on IPRs by the industrial nations, many did not expand their own levels of protection until technological development was well under way. Canada, for example, allowed patents for pharmaceuticals beginning only in 1987. Schiff (1991) found no evidence that Holland or Switzerland was hampered economically during their patentless years (1869–1912 and 1850–1907, respectively). Current evidence suggests that strong IPRs are neither a necessary nor a sufficient condition for R&D investment, at least in certain sectors. Grief (1987), however, showed that in Germany R&D investments and patent applications were closely correlated, suggesting a role for patents in stimulating investments. IPRs are typically ranked low by corporate executives as a stimulant to private R&D investments. In a review of a number of empirical studies conducted since the 1960s, Mazzoleni and Nelson (1997) conclude that in most industries (pharmaceuticals excepted) patents were not an important incentive for investing in R&D. These results are in general agreement with company surveys of the importance of IPR to investment decisions (Nogues 1990).

Pharmaceuticals may be an exception, mainly because it is a high-profit industry for which effective IPR protection is an important stimulus to R&D and technology transfer. A more likely explanation is the regulated nature of pharmaceuticals. Most of the estimated US$250 million cost of bringing a successful product to market is associated with safety and efficacy trials, a part of the regulatory process. Competitors could potentially save two-thirds of that amount in registering another version of a previously approved drug, meaning the lead firm could not compete. IPR protection provides a period free from direct competition during which the high initial investment can be recovered. IPRs are, therefore, often considered to be particularly relevant for regulated products, such as pharmaceuticals and, arguably, agricultural biotechnology. There is to date no direct evidence of the effects of IPRs on pharmaceutical R&D investment. Canada's experience, for example, has yet to be documented. Deolalikar and Evenson (1990), however, found that pharmaceutical R&D in India fell by 40 percent from 1964–70 to 1980–81, following a weakening of patent protection in 1970.

An important research question in the case of agricultural IPRs is: what is the net benefit for developing countries, of different sizes and stages of development, in moving from an essentially non IP protection regime (often with legislation that is not effectively enforced) to one that is enforced?

Evidence is limited for agriculture in developing countries since only Argentina and Mexico had transgenic crops grown in 1997. There has not yet been an analysis of the experiences (Union of Concerned Scientists 1997). A study of PBR in Argentina suggests the importance of IPR protection because "PBR enforcement seems to have played a role in sales and R&D expenditures of domestic companies involved in markets for self-pollinating varieties" (Jaffe and van Wijk 1995, 48).

In agriculture, particularly in developing countries, effects of IPRs on R&D investments are likely to be specific to the crops and the structure of national agricultural systems, and thus difficult to predict based on experiences elsewhere. For example, there is a greater likelihood of considerable private expenditures on soybean in Brazil, which is grown by large farmers and exported, than on maize and root crops grown by low-income farmers in the Northeast. These differences can be explained by the potential commercial opportunities in these different crops. It is quite possible, however, that many of the IPR protected technologies would be scale neutral in application on farmers' fields, particularly those involving biological manipulations. These, therefore, would be potentially available to small farmers provided they have access to information, finance, and markets. An important question for the future is: what role, if any, would the large multinational seed companies play in ensuring access by the poor to improved varieties?

There are also important differences in the way the private sector might view large and small countries (for example, in terms of the scientific infrastructure, including the availability and the cost of skilled labor to conduct research and technology transfer, and the ability to copy technological advances and hence the extent to which countries are seen as partners or clients, rather than competitors). These factors in turn may influence the incentive the private multinational corporations would have to invest in research.

Although most researchers do not include country size among their explanatory variables for the impact of IPRs, Maskus and Penunbarti (1995) looked at country size in relation to IPR-induced trade flows. They found that whereas IPRs have a positive impact on trade in small and large countries, the impact is stronger in large countries. *Important questions: to what extent do IPRs affect the productive activities of small, medium, and large farmers? in what ways are IPRs likely to be exercised among small, medium, and large countries?*

Technology Development and Transfer

Developing countries tend to be significant importers of technology from industrial countries (Braga, Fink, and Sepulveda 1998). In the area of industrial property (patents, trademarks, industrial designs), less than 5 percent of the 2.1 million patents granted worldwide in 1993 went to developing countries (Braga, Fink, and Sepulveda 1998). In contrast more than 80 percent of patents granted in developing countries were filed by residents of industrial countries. Grief (1987, p. 207) notes that foreign holdings of patents are higher in smaller industrial countries (Belgium, Canada) than in many developing nations. Indeed, the per capita proportion of foreign-owned patents rises with increasing per capita GDP.

Some use this evidence optimistically to argue that, on the assumption of equal distribution of inventive minds across the globe, there is a large natural resource to be tapped in many developing countries awaiting stronger IPRs to reach their potential. They also argue that in countries with weak intellectual property protection, inventors have been discouraged from filing patent applications (Sherwood, Scartezini, and Siemsen

1999). There are concerns in developing countries that the IPR offices may be sources of leaks.

Price Effects of IPRs

Patents create a limited right to exclude others from using an invention, but not from the market served by the invention. The patent owner will often be in a position, at least temporarily, to extract higher rents. Price effects of IPRs have been intensely debated, although the underlying issue relates more to the cost/benefit ratios of new technologies relative to the existing ones, and the extent to which markets in information, finance, and commodities enable all (including small) farmers to access the new technologies.

The limited evidence drawn from countries with PBRs or with private sector hybrid seed (which can be considered as having a form of biological protection, as hybrids do not reproduce true-to-form and must be replaced annually) indicates somewhat higher prices with IPRs. This would be expected and necessary to recover private R&D expenses, but there appears to be little evidence of excessively high prices with agricultural inputs. Evidence includes

- USA: PBR prices are above unprotected variety levels but involve a rent sharing between farmers and breeders (Butler and Marion 1985). In general, farmers under-invest in purchasing seed and suffer yield and income losses compared to more frequent seed purchases (Knudson and Hansen 1991). Certificates of PBRs were associated with a 2.3 percent higher price for soybean seed in New York State (Lesser 1994).
- Argentina: the additional costs of licensed protected varieties were not passed on to farmers even though PBR legislation requires seed dealers to pay royalties and taxes (Jaffe and van Wijk 1995).
- Brazil: prices for high-yielding maize varieties were related to yield potential and were significantly higher than public variety prices (Garcia 1998).
- Mexico: public maize varieties are significantly lower priced because of the absence of a need to cover breeding costs and the target of small-scale as opposed to commercial-size farmers (Aquino 1998).

A finding of interest to developing countries is that even with the new agricultural biotechnology crops for which there are few suppliers (Monsanto may have upwards of a 77 percent market share worldwide), the sharing of rents between users and seed suppliers is within the long-prevailing U.S. 50/50 pattern (Flack-Zapeda and Traxler 1998). More such studies are needed, particularly in developing countries, before there is a strong basis for anticipating the effects of potential private sector investments on seed prices. Despite this (clearly limited) evidence, many countries have justifiable concerns over the undue extension of IPR monopoly power because they lack effective antitrust protection.

Undue extensions can include such acts as using a patented device to require other products or services be purchased from the patent holder (tying arrangements or full line forcing). These issues have not been examined in depth since the mid 1970s when compulsory licenses were recognized as one available mechanism to control patent abuses (UNCTAD 1975). Compulsory licenses permit a government to grant a use permit for a patent without the permission of the patent owner. Available evidence indicates that compulsory licenses are rarely granted, but the issue remains open as to whether their existence provides a controlling factor in markets, which largely removes the need of enforcement (see Roffe 1974; UNCTAD 1975).

Under the TRIPs agreement (Articles 30 and 31), "[M]embers may provide limited exceptions to the exclusive rights conferred" subject to conditions of individual merits of requests, substantial efforts to achieve agreement with the owner, nonexclusivity, and subject to judicial review. When the request applies to dependent patents, the request must be for "an important technical advance of considerable economic significance." (A dependent patent includes, among other things, a patent for an improvement to a patented product or process. The owner of the dependent invention would require the permission of the underlying patent holder to commercialize the improvement.) These issues are quite complex and need to be interpreted in conjunction with the relevant parts of the Paris Convention (Articles 5 and 10; see Reichman 1993; Heinemann 1996; Straus 1996). An example of a use of compulsory-licensing authority is the revised

Canadian Patent Act (Chap. 33), which provides for a Board to receive pricing information on patented medicines along with the authority to request a price reduction, among other steps.

Areas for Further Research

We have identified a number of areas in which the current level of understanding of the roles and effects of IPRs are insufficient to provide clear guidance to countries planning changes in the form and implementation of IPRs. In some cases the data needed for the analysis have been available for some time. In other cases particularly with regard to commercial agricultural biotechnology products and the operation of PBR in a number of countries, field data have become available only recently, making in-depth analysis now feasible. Many of the areas raise important policy and institutional issues, including:

- Effects of PBR in developing countries on product prices, availability, adaptation for different farm types, and agronomic merit of protected varieties
- Extent of any abuse of IPRs, and uses and consequences of compulsory licensing legislation in overcoming any such problems
- Effects of IPRs on access to state-of-the-art technologies, particularly for agricultural biotechnology products, to include a more complete understanding of the transfer strategies of multinational firms
- Study of the nature and effect of broad patent claims in agricultural biotechnology
- Changes in public and private investments in national R&D programs following strengthening of IPR systems
- Equality of access to IPR systems by national entities, including small firms and the public sector.

The World Bank, with the active participation of its borrowers and member countries, is in a strong position to be a catalyst in establishing an international research program to provide additional empirical information on these and related issues.

Strategies to Strengthen IPRs in Developing Countries

It is a long and unknown path between adopting legislation and rendering effective protection to intellectual property. This uncertainty presents a number of challenges to domestic and international negotiators on intellectual property protection. Beyond the minimum TRIPs and the CBD-related requirements, countries must decide what level of IPR protection is appropriate, comparing the available types of IPRs, and exploring the various legal possibilities available. The appropriate level of national legislation and the practical implementation issues are discussed below.

Selecting Optimal IPR Protection

Sherwood (1997) distinguishes between three levels of IPR protection: (1) nonrobust, (2) TRIPs-compatible, and (3) investment-stimulating/robust. The final assumes the judicial systems work effectively. He argues that at nonrobust levels of protection countries are able to encourage sales and distribution, assembly, and component manufacture. However, protection that stimulates private investment in higher technological activities appears to be viable only at a level somewhat above the protection offered by the TRIPs agreement. At a TRIPs-compatible level, protection will support imports, as well as creative expression and somewhat more ambitious industrial activity. However, only at levels of protection above the TRIPs requirement is local research stimulated, and the development of more sophisticated technology likely to be supported by private investors.

Sherwood hypothesizes that at nonrobust levels of protection, investment will be of short duration, whereas at robust protection levels it will be more durable. Perhaps more vital for sustained economic development, minimal employee training is attempted under conditions of nonrobust protection, whereas training in high levels of technology will be stimulated at robust levels of protection, elevating the human capital of a country at private expense. Also at robust levels of protection, risk capital is mobilized, university research results find commercial application, and higher science is likely to be applied to agriculture, among other areas. Once robust IPRs are in place, these developments begin to take place naturally, without the need for government interventions. However, these conclusions are based on limited empirical analysis and must be interpreted with caution until a more solid database is established.

Sherwood (1997) goes on to contend that trade secret protection is perhaps the least studied form of protection with importance for agricultural technology. A trade secret is information of value to the holder for which an effort is made to prevent unauthorized access. A major benefit of trade secrets is that there is no limitation to duration so long as secrecy is maintained. However, protection (unlike that for patents) does not prevent the discovery of the secret by any sort of "reverse engineering." In other words the secret must be undiscoverable after a product is marketed and subjected to thorough analysis.

One IPR option well suited to developing countries may be the petty patent, also known as the utility model. This form of protection in essence allows a smaller scope and duration of protection for more limited inventions. Thus it is suited to minor improvements or adaptations that are within the purview of local artisans in developing countries (Evenson, Evenson, and Putnam 1987; Ojwang 1989). Countries not offering that option may wish to consider it. However, because many national laws limit petty patents to the mechanical arts, such patents would not be applicable to all classes of inventions, including those associated with living materials.

Patent legislation allows scope for choice among alternatives. At one level is the distinction between registration and examination systems. Examination systems include a detailed search of the available literature so that successful applications have a presumption of validity. Registration systems for their part involve only the recording of applications; any determination of validity is not undertaken until there is a challenge. Registration systems are inexpensive and simple to operate, but place a heavy reliance on the court system. Conversely, examination systems are complex and costly to operate, but provide more clarity in the rights bestowed. Within examination systems there are alternative operating procedures, such as the use of a national examination, examination by another major national office, through the Patent Cooperation Treaty (PCT), by a private search firm (for example, International Patent Documentation Center (INPADOC) in Vienna), or through a regional IPR office.

Countries can choose between patents or PBRs (different options for implementing plant breeders' rights are presented below to protect plant varieties). Under the PBR system, governmental authorities often leave farmers the freedom to use their own harvested material of protected varieties for the next production cycle on their farms. This privilege is referred to as the *farmers' privilege* but it has become "optional" in the UPOV Act of 1991, which leaves the matter to national jurisdiction. The plant biotechnology industry is increasingly resorting to patent law because this offers stronger protection than PBR. However, even though the role of the private sector is increasing, varietal development in developing countries is still predominantly the domain of public institutions. This system relies on the international free exchange of plant genetic material, and on-farm seed saving practices. It may face restrictions when plant varieties are protected under patent or PBR law.

The examination vs. registration choice also exists for PBR. The U.S. Plant Variety Protection Office, for example, operates as a modified registration system compared to European countries, which grow out applicant varieties and test for differences in the claimed distinct attributes (Lesser 1987). Examination provides clearer rights and some buyer protection, but is costly and slow to operate (two-year trials are common), so there are clear tradeoffs between the two approaches. In some countries, concerns over the examining office's ability to prevent unauthorized access to varieties under evaluation is a disincentive to using variety registration.

In the past countries have often followed a hierarchy of systems. The U.S. Patent and Trademark Office, for example, initially operated a registration system, switching in later years to examination. Conversely, for many years African countries such as Kenya operated under a "re-registration" system, whereby a Kenyan patent could be approved only if one had been issued in the United Kingdom (see Ojwang 1989). Re-registration, while reducing the responsibilities of the national patent office, placed a heavy burden on developing country applicants, who must satisfy and pay for a patent in an industrial country where there may be no intention to market the protected product. This option, while attractive from an administrative perspective, may not be practical for developing countries unless different fee structures and patentability standards are established for developing country nationals. This explains why separate national patent systems have emerged in the post-colonial period.

Access Legislation and Benefit Sharing

Paralleling these developments in traditional IPR law, developing countries are concurrently adopting legislation controlling access to genetic resources, and providing a basis for benefit sharing. The CBD specifies that genetic resources are the "sovereign right" of nations to determine access, but rather than the right being automatically bestowed, parties "shall take legislative, administrative or policy measures" to achieve equitable sharing and access (Article 15). To date three entities have adopted comprehensive access legislation (UNEP/CBD/COP/4/23, Feb. 1998): the Philippines, Andean Pact member nations (Bolivia, Colombia, Ecuador, Peru, and Venezuela), and Brazil, Costa Rica, and India have legislation in an advanced draft form.

Access legislation is closely related to the issue of "biopiracy." In the recent dispute relating to the patenting of a new hybrid strain of basmati rice by an American company, considered by many in India to be a form of "biopiracy," the chief executive officer of that company argued India should have obtained a United States trademark on the name "basmati" if it was considered a valuable property warranting protection. That stance suggests a preemptive behavior in which owners of valued properties must take proactive steps to protect their assets, rather than relying on deference to informally recognized prior claims. Proactive positions will, in turn, require the availability of appropriate protection mechanisms. The basmati rice dispute is an example of the intricate way in which various IPR issues relating to plant genetic resources can become interwoven: for example, the appropriation of genetic materials that originated in other countries in possible violation of the Biodiversity Treaty, the patentability of hybridized rice strains, the registration of common terms (in Hindi) as trademarks in other countries, the passing-off of the place of origin, and possibly other issues. The basmati rice case is discussed in Watal (this volume).

Also involved are issues of indigenous and traditional rights (CBD Article 8(j)), where effective forms of legal protection are sought (Decision IV/9). Many argue that traditional forms of IPR are not suited to protect indigenous knowledge. The limitations are: (a) conceptual—traditional knowledge is often not embodied in a product, which is the form protected by IPR; (b) procedural—patents require a showing of utility or industrial application that could not be provided for many unevaluated species; and (c) practical—the costs of protection are excessive for the great bulk of materials that will not have any commercial value. Hence, some new systems will be needed, which could include a role for the World Bank to study these issues leading to alternative solutions. An overview of current proposals (UNEP/CBD/COP/4/10 and UNEP/CBD/COP/4/23) is summarized in Lesser (1998a), where a proposal for a system based on revealed knowledge is also presented.

Options for Implementing TRIPs

The most pressing requirement for many countries is the need to comply with the commitments made under TRIPs. The general requirements are quite specific, but within them are a number of options that might be considered. Critically, TRIPs agreements allow countries to select "an effective *sui generis*" form of PBR as opposed to patents for plants, a choice many countries prefer. Within this context, countries retain at least three choices (Leskien and Flitner 1997; Seiler 1998):

1. Join the international PBR convention (UPOV).
2. Develop a related national version of PBR without acceding to UPOV.
3. Devise a distinct sui generis applicable to national needs.

Each choice has significant potential effects. The choice of the 1991 UPOV Act, for example, requires protection to be extended to all plant genera and species within a 10-year period. A national law permits more flexibility but does not ensure reciprocal rights in other UPOV member states, which is a major drawback. Countries may wish additional information on options prior to committing to a particular form of PBR legislation.

Practical Considerations

The actual implementation of the new IPR legislation may prove a more complex issue than the passage of the enabling legislation. It takes time to develop a good environment, and time is becoming an important constraint for countries that have to comply with TRIPs obligations. Some of the more practical considerations countries face are discussed below.

Regional or Global

Countries can operate through a centralized multi-country office. The best established of these is the European Patent Office (EPO), which performs search and examination functions for the EU participating nations (as well as for the 95 countries that are members of the Patent Cooperation Treaty; see below). Although applicants can simultaneously apply for multinational grants, the grant is a "bundle" of national patents, rather than a single grant for the designated member countries. More integrated, but on a far smaller scale, is the agreement between Liechtenstein and Switzerland under which the former processes the patent applications for the latter. In southern Africa, there are two long-standing regional patent offices, the Organisation Africaine de la Propriete Intellectuelle (OAPI), and the African Regional Intellectual Property Organization (ARIPO), which could serve as models for other regions. However, their activities and effectiveness are not well documented (for example, in 1995 OAPI granted 31 patents and ARIPO 68, which suggests that these offices are not very active (WIPO data)). The EU nations have recently implemented a centralized PBR examination system under which national PBR offices specialize in testing specified crop species for all member states.

On the convention level is the PCT, which permits member states to request international searches and international preliminary examinations by designated offices. The latter can complete most or all of the work typically required of a national patent office. However, although the PCT is quite active, receiving 54,422 applications in 1997 corresponding to more than 3 million national applications (WIPO 1998), its principal use has been to delay by up to 18 months the time when applicants must pay national filing and translation fees (see below).

The PCT has the benefit of being an existing, ongoing agreement. Regional patent offices require enabling legislation, as in the case of the EPO, for harmonized national patent laws. Even then, the EU states have maintained national offices, albeit the number of patents granted has been reduced considerably following the growth of the EPO. The PCT is already extensively used, with 54,000 applications corresponding to over three million national applications in 1997 (WIPO 1998). Additional uses of the PCT, administered by WIPO, can perhaps best be supported by that organization.

However, the PCT, while effectively postponing expenditures, does not avoid them. If an inventor has not succeeded in licensing the invention by the end of 30 months from the first application, full application and translation fees will fall due. Abbott (1993) estimates that translation costs double the costs of patents in Europe compared to the United States, although recent modifications in European translation requirements reduce those costs somewhat. The delay benefits, although real, are at the margin, because the grace period under the Paris Convention allows the applicant 12 months from first application to apply in other member states (Article 4). This means that the full search and examination costs for all countries in which a patent is sought be paid 30 months following the date of first application. Perhaps most significant is the fact that the major costs of complex applications like those for agricultural biotechnology products are for attorney fees and not for processing costs. The attorney fees must be paid at the time of the initial application. For the United States the costs for an agricultural biotechnology patent are about US$25,000, whereas protection across OECD costs upwards of US$100,000 (Walter Haeussler, Director, Cornell Research Foundation, May 1998, pers. comm.). Meller (1998) places the world costs for a single invention at US$500,000. A more viable strategy for the impecunious inventor may be to apply nationally, and seek to license the invention within the 12-month grace period, with the requirement that the licensee is responsible for the costs of patenting in other countries.

Cost of Patent Acquisition

From the relatively small number of patent applications in developing countries, it can be assumed that there are many more inventions for which no application is ever filed. Among the most important reasons for this phenomenon is the cost of patent acquisition. In Brazil (this section draws heavily on the Sherwood, Scartezini, and Siemsen (1999) paper "Promotion of Inventiveness in Developing Countries Through a More Advanced Patent Administration") and elsewhere, patent agents are familiar with innumerable clients who have approached them seeking advice regarding their inventions, but who subsequently retreated without applying upon learning of the costs. Equally, if not more, important than the burden of patent acquisition costs is the timing of those costs when they compete with capital demands for product development. This suggests that patent-related costs should be postponed as long as possible, preferably to near the moment of commercialization.

Patent acquisition costs include, in part, official fees for search, technical examination, and translation. One way to reduce these costs would be to draw more strongly on the global system through the PCT (see above). If fully used, national costs of patent examination could be greatly reduced; applicant costs will depend on the national and PCT fee structures. For applicants, the PCT provides the additional advantage of postponing for up to 18 months the cost of obtaining multiple patents.

At the national level two concepts have been proposed to moderate the level and timing of patent acquisition costs, namely rapid patents and patents-by-reference.

Rapid patents. This concept consists of extending authority to the national patent office to grant patents immediately after an examination of the patent application formalities, but without prior technical examination. Technical examination could be requested at a time selected by the patent holder—before or immediately after the grant of the patent, or postponed indefinitely. This approach gives the applicant a degree of flexibility over time-related patent acquisition costs, and thus gives the applicant a better chance to develop the invention. At the same time this approach reduces public administration expenses associated with the nine out of ten patents that will never be commercialized.

Patents-by-reference. Many countries face considerable difficulty in providing high-quality technical examination for all patent applications received. For them it has become the practice to wait for the results of patent examinations in some of the major international examination centers, essentially those designated by the PCT. This being so, some contend that reliance on the PCT system could be taken a step further, if countries were to grant patents more or less automatically, and without local examination, once one of the major PCT-designated examination centers grants a patent for an invention.

A patent-by-reference system could be created unilaterally by a developing country, without major changes in the current law. It would relieve developing countries of the burden of patent examination, and would offer a relatively low-cost way for developing country inventors to obtain carefully examined patents in as many countries as adopt such a system, and would impose modest fees. This could provide a powerful incentive for inventiveness in developing countries, and would free up public resources to create electronic information services to access world knowledge on agricultural and other technology.

Those who take positions counter to Sherwood's could argue that "rapid patents" are similar to registration systems (see above). Registration systems are also vulnerable to a situation under which there are many patents on the books for which the scope and indeed validity is not clear; indeed, if validity were in question the owner would be unlikely to request examination. Unclear property rights typically chills research as no one wishes to invest in a product when the rights to its use are unclear. Moreover, if an unexamined patent is challenged in court, a considerable technical burden is placed on that body. Since the courts are considered one of the weak components in developing country IPR implementation, the use of rapid patents could exacerbate the overall inefficiency of the system.

Patents-by-reference closely resemble the colonial era-based re-registration system that has been replaced with national systems. Patents-by-reference can indeed reduce governmental administrative costs, and provide a patent of carefully considered merit. However, that option is now available to developing country inventors who apply first in a country with a leading patent office. The reliance on industrial country patent offices to make a decision on patentability can, however, have two effects. First, it establishes a very high technological standard on the developing country inventor. A national system could adopt a more limited standard if that were appropriate. Second, national governments might be at variance with the patentability standards established by some industrial country offices. One can hardly imagine the protests that would have been raised if India, under a patent-by-reference system, were committed to grant the turmeric and basmati rice patents approved by the U.S. Patent and Trademark Office!

Overall, patentability standards are a significant aspect of national industrial policy. Careful consideration needs to be given before ceding it to another country under a patent-by-reference system.

Operation and Enforcement

TRIPs has several enforcement-related obligations within its provisions. Some observers propose the use of a specialized court for IPR matters, but that is not mandated by TRIPs (Article 41(5)). The requisite technical competency of a court is sometimes used to justify specialization, but adequate expert advice can make up for the lack of technical knowledge of the justices. Indeed, no panel can be technically knowledgeable in all subject areas likely to appear before it. Effective implementation typically entails a complex and routinely tested web of efficient domestic institutions. In addition to the judiciary system, which is elaborated on further below, national patent offices are the most obvious example. One practical question that arises in the context of PBR is whether these should be handled in a separate office. In smaller countries it might be expensive to set up an independent system when much of the expertise is in the research system (for example, for varietal testing). However, Sherwood (1997) points out that contracted examination or testing for a fixed fee has some drawbacks that

countries should try to avoid. In Zimbabwe for example, certification is the responsibility of the Seed Services which also produce and market their own varieties and hence are seen as potentially biased against other seed producers (Rusike 1998). Canada has a system under which PBR applicants are responsible for managing and funding variety testing under the supervision of the PBR office.

More fundamental, and problematic, is the general functioning of the court processes in adjudicating IPR law. Some court systems are overburdened to the point that cases are delayed for years—one level of the Indian court system is said to have a case backlog of over 1 million. Others do not operate in a transparent fashion, reducing confidence in the rendering of a "fair and equitable" judgment. These are fundamental issues for IPR since it can be presumed that there is no effective protection, and hence no business response, unless and until the rights are enforceable (Jaffe and van Wijk 1995).

Much IPR is litigated in civil cases. Indeed, most cases are brought by the allegedly damaged party seeking compensation. This means that governmental costs are low, but the appropriate courts and practices must be available.

Building National Consensus

IPR and related matters are politically controversial and emotional. In the past, thousands of Indian farmers have marched in protest of PBRs. Governments have taken positions different from small-scale farmers and NGOs, whereas the private sector may represent a distinct third position. The numerous bases for the divisions in perspective include
- Views regarding the sanctity of life
- Equity within and between countries
- Failure to resolve past differences on compensation for genetic resources
- Divisions between national governments and indigenous peoples on access and use of natural resources
- Misunderstandings, such as a failure to understand that PBRs do not affect previously existing uses of protected materials.

Differences in perspectives delay the achievement of national and international consensus and, further, make the implementation of the CBD difficult. Certainly one of the ongoing areas of discussion is that of Farmers' Rights.

The term "Farmers' Rights" arose from the FAO's Revised Undertaking for Genetic Resources (Resolution 5/89), where they were defined as "rights arising from the past, present, and future contributions of farmers in conserving, improving, and making available plant genetic resources." In 1991 an international fund was established for implementing Farmers' Rights, but has yet to receive any substantive contributions (Resolution 3/91, Annex 3).

Farmers' Rights have been discussed at every Conference of the Parties of the CBD to date, but with no consensus on how to proceed. Complicating the discussions are the multiple interpretations of Farmers' Rights. The term grew out of an effort to promote acceptance by the international community of its common responsibilities towards a Global Plan of Action to develop and conserve Plant Genetic Resources for Food and Agriculture (PGRFA), and redress perceived inequities in the benefits flowing from the provision and use of PGRFA, not as an alternative form of IPR (Fowler 1998). Distinguishing between political and legal rights has, however, proven elusive, as has even the basis for measuring the economic value of farmers' contributions, whether it be their economic value or proportion of genetic material contributed. India, in drafts of a national PBR system (see Damania 1996; also Watal 1996), recognizes Farmers' Rights internally with a proposed tax on seed sales, with proceeds to be directed to source communities when identified, otherwise to conservation. In practice, this operates like a national tax and is quite distinct from the concept of the "North" compensating the "South" for the use of genetic resources, which seems to underlie some of the discussions.

Some have treated Farmers' Rights as a form of quasi-IPR, but they are distinct in several key ways. IPRs provide specific market success-based incentives for commercialization of products and processes. Farmers' Rights are to compensate for "past and present" conservation and development not directly associated with the market value for each contribution. Farmers' Rights do recognize "future" contributions as well, but again not in terms of individual market rewards. However, existing IPR legislation is not well suited to protect wild relatives and traditional varieties, which typically lack the stability and uniformity to qualify for PBR. Hence, some new system would seem to be required.

Conversely, the use value of most accessions held *ex situ* in international collections (the "average" accession) is estimated to be small and the transaction costs high, so that raising significant sums through sale is also unlikely (see Perrings 1995; Lesser 1997). Although the vast bulk of requests for materials from the international collections comes from national programs in developing countries (data in Lesser 1997), there is considerable quantitative evidence of vast benefits of the CGIAR materials to industrial countries. The private sector currently requests few accessions, in part due to the use of their proprietary collection, and in part due to the CGIAR policy discouraging IPRs on CGIAR-supplied materials without ensuring free access by developing countries to resultant varieties. That policy is currently under review (see below). Overall, the paying base from which to generate funds, even if a system could be implemented, seems limited. Thus the matter of Farmers' Rights remains controversial with no resolution apparent at this time.

IPR Implications for Public Sector Research

Public sector research can be directly affected by IPR in several ways:

- The opportunity to patent a discovery is inadvertently lost when it is publicly revealed (Many countries operate under the so-called absolute novelty system where any public revealing of the invention prior to the first application destroys the opportunity to patent.).
- A research contract provides for limitations in the publication of results.
- An institution's IPR policy gives it control over the use of employees' inventions, including the right to grant exclusive licenses.
- The broad scope of some recent IPRs in the area of plants and agriculture means a scientist's research could be infringing IPRs, leading to possible legal action.
- The development of products using materials provided under research MTA has in some cases resulted in an inability of the researcher

to subsequently secure a commercialization license, rendering the prior work useless and raising questions about how to proceed in the future.

Need for Training

Most scientists, science administrators, and institutions are unfamiliar with IPRs and their specific roles and implementation. This calls for more and better training, information dissemination, public education, and curriculum development in colleges. As resources for research in developing countries are offered through competitive grant programs that reward public-private partnerships, important issues faced by scientists include: which partners (scientists from public institutions and private firms contracting research) should receive what benefits and how costs should be shared. Not only do scientists lack exposure to the legal and financial issues, but also to the ethical issues involved in the establishment of public-private partnerships.

In developing countries unlike in their advanced industrial counterparts, training is not sufficiently based on the particular needs of the individual and institutional clients. Because training lacks the user's perspectives in the specific countries, there is often a dilemma: those with particular designated responsibilities for IPRs often lack the necessary human capital, while those few trained are typically not trained to address specific problems in their own countries, despite advanced degrees from recognized western universities. Once trained they are rarely deployed in their home countries to advance intellectual property protection. Addressing training and effective utilization of trained scientists and science managers is a major and urgent need.

Learning from CGIAR-IPR Experience

The private sector response to the CGIAR policy of not seeking IPR protection has important lessons for the public research systems of developing countries. The private sector considers the CGIAR policy to be ineffective for expanding private sector activities and mobilizing support. Not seeking IPR rights permits private firms to protect improvements over which the interna-

tional Centers have no control. Conversely, ownership of IPR permits considerable flexibility in use—the open licensing to one group (small-scale farmers) with fees required from another (multinationals, commodity crops). An IPR portfolio in the Centers could also strengthen negotiations with the private sector at a time when there is often more willingness to cross-license than fee-license protected materials (Barton 1998).

Centers are now responding to this challenge in a variety of ways. For example, in a CIMMYT/ORSTOM (Centro Internacional de Mejoramiento de Maiz y Trigo/Institut Francais de Recherche Scientifique pour le Developpement en Cooperation) collaboration on breeding of apomictics through conventional and molecular strategies, the evaluation is aimed at assessing the potential economic impact of apomixis and the assessment of biological risks. Because apomictic maize will not change characteristics from generation to generation, farmers can purchase seed only once. But for just the same reason, apomixis can enhance the genetic uniformity of crops. ORSTOM applied for a patent in its and CIMMYT's name as a defensive measure to control use. However, for such an important development more investment will be required to achieve broad coverage. Hence, the intent is to seek a third party to undertake these expenses. CIMMYT and ORSTOM can provide the third party with a license to use the base technology in the industrial world, requesting rights to make it available on a nonexclusive basis in developing countries. This arrangement provides benefits for the third party, for CIMMYT and ORSTOM, and the developing country farmers. The NGO community has come to hail such agreements as the way of the future.

The CIP (Centro Internacional de la Papa) Board of Trustees recently passed a resolution on the "Sharing of Benefits of CIP-Generated Technologies through Intellectual Property Rights," in which it recommends that the Center strengthen its IPR policy and use any income generated to implement the Global Plan of Action for the Conservation and Sustainable Utilization of Plant Genetic Resources for Food and Agriculture. Furthermore, at the Mid-Term Meeting of the CGIAR in Brazil in May 1998, a Centers' position statement on genetic resources, biotechnology, and intellectual property rights emerged. It summarizes the current practice of Centers with regard to ethical principles, agreements with FAO, standard agreement forms for germplasm acquisition and material transfer, guiding principles on intellectual property rights, interaction with the private sector, and an agreed position statement on biotechnology.

Monitoring the Germplasm Flow

Neither FAO nor the CGIAR currently have adequate resources to monitor germplasm flows from their collections. In the case of *ex situ* materials, tracking and litigation costs are estimated at US$40–500+ per accession depending on the methods used. With 800,000 accessions distributed annually, the cost could be US$32–400 million annually (Lesser 1997). Not only is the cost exorbitant, but financial returns from the sale of materials may be negligible. There is also the issue of the capacity to defend claims. In a recent case cited by the Rural Advancement Foundation International (RAFI) following a demand from ICRISAT, Australia abandoned PBR applications based on materials acquired from a designated "in trust" collection. While such public actions receive considerable attention, important questions are: Who should monitor and how should germplasm use overall be monitored? In particular, should nongovernmental organizations (NGOs) continue to play the role they are playing now or are other approaches needed?

Sharing Collaborative Models

There is a need for a range of models of open and flexible arrangements between the public and private sectors. Monsanto typifies one approach to biotechnology. A widely shared view including in the CGIAR centers is that approach emphasizes significant central control through ownership and contract (the 1997 Monsanto annual report lists 16 acquisitions and joint ventures over a 14 month period; in 1998 Monsanto acquired Delta and Pine Land and DeKalb, two major seed companies). That is an effective business strategy, but one that raises questions about the role of the public sector. In contrast, Novartis has adopted a more open process regarding genetic mapping of the rice genome. Operating es-

sentially to assume the Rockefeller role in broadly supporting rice research, Novartis is funding public sector basic research and making the results public. The right is retained for Novartis (and any other firm for that matter) to protect derivative applications, but the public sector contribution is clear and basic information remains in the public sector. That model should be observed closely with the intent of finding incentives to replicate it in other cases.

The concept of reserving basic materials for public use while commercializing applications can be applied in other cases. It might help resolve the long-standing impasse between genetic resource providers and users where the former expects payment but the latter sees the materials as either in the public domain or as of limited value at the individual accession level (Collins and Petit 1998). An alternative approach is to reserve the materials (with available passport data) for open public access, but to sell any characterizing information. That approach, while sidestepping the complex issue of genetic resource ownership, also provides an incentive for programs to invest in characterizing materials that have little commercial value if users must perform the screening. This would create a value-added process with the potential of benefiting users and suppliers (for further development of this approach see Lesser 1998a).

There is also an urgent need for skilled negotiators, and better information on standard contract terms. These needs, along with possible responses, are further developed in Lesser and Krattiger (1994). Presently, no international entity is responsible for these needs, a gap that could be addressed by the World Bank. One approach could be an independent for-fee agent that provides negotiating assistance on request, and, as a participant, could collect information on license terms to be made available in a nonattributable form. Contract terms are not now made public, so that inexperienced negotiators have difficulty differentiating between good and bad agreements.

There is probably also a need to identify standard (or model) agreements for licensing component technologies, in much the same way that the World Bank-financed projects have over many years developed some standardization in the legal documentation of projects. The present approach of "zero based" negotiation for each user is costly in time, meaning that many less significant agreements are never completed. The World Bank could be instrumental in establishing an expert panel of public and private sector representatives to identify baseline agreements. These would serve only to facilitate reaching final agreements, not as compulsory contracts, or even as a single blueprint.

Ensuring Public Goods Priorities

The international and national public systems need to guard themselves against the danger of research priorities being set by private company research funding, unless such research has a clear public goods content. This issue arises increasingly often in many national systems of developing countries, as finance ministries encourage the collection of revenues from the commercialization of their own research. These risks have been noted in the NARS of Brazil and China, while at Cornell University, a long term project on coffee production was undertaken in part because of the availability of funds when locally grown crops were not so supported.

Establishing a CGIAR-World Bank Facility?

At the CGIAR Mid-Term Meeting in Brasilia in May 1998 a decision was taken to create a central advisory facility on legal matters with access to appropriate expertise, to assist the CGIAR centers in addressing IPR issues, a decision endorsed by the CGIAR system-wide review. This facility's scope, content, location, and management responsibility remained to be determined as of the finalization of this volume. The World Bank and the CGIAR could work in partnership, and on a cost-sharing basis, to ensure that similar expertise and experience would be available to developing countries on a worldwide basis, possibly through the use of their loans and credits provided by the World Bank for accessing expertise. To the extent that the CGIAR's impact on poverty alleviation depends on the rapidly evolving laws and their implementation in developing countries, a joint facility might help achieve greater correspondence between the national and international systems. But this may not be sufficient. The World Bank will need core capacity at the center in its Rural Development Family to deal with the issues involved in the design and implementation of its lending operations as they pertain to IPRs and the use of biotechnology more generally. These implications are explored in the final chapter of this volume.

References

Abbott, A. 1993. "Monoglot Filing Urged for European Patents." *Nature* 364: 3.

Agarwal, A., and S. Narain. 1996. "Pirates in the Garden of India." *New Scientist* 26/10/96, 14–15.

Aquino, P. 1998. "Mexico." In M. L. Morris, ed., *Maize Seed Industries in Developing Countries*. London: Lynne Rienner Publishers, Inc.

Barton, J. H. 1998. "The Impact of Contemporary Patent Law on Plant Biotechnology Research." In S. A. Eberhart, H. L. Shands, W. Collins, and R. L. Lower, eds., *IPR III—Global Genetic Resources: Access and Property Rights*. Madison, Wis.: Crop Science Society of America.

Beier, F.-K., and G. Schricker, eds. 1996. *From GATT to TRIPs—An Agreement on Trade-Related Aspects of Intellectual Property Rights*, vol. 18. Weinheim: VCH.

Braga, C. A. Primo. 1995. "Trade-Related Intellectual Property Issues: The Uruguay Round Agreement and Its Economic Implications." In W. Martin and L. A. Winters, eds., *The Uruguay Round and the Developing Economies*. World Bank Discussion Paper 307. Washington, D.C.

Braga, C. A. Primo, C. Fink, and C. P. Sepulveda. 1998. "Intellectual Property Rights and Economic Development." World Bank, background paper to the *World Development Report 1998*, Washington, D.C.

Butler, L. J., and B. W. Marion. 1985. "The Impacts of Patent Protection on the U.S. Seed Industry and Public Plant Breeding." U. Wisconsin, Ag. Exp. Station, NC Project 117, Monograph 16.

Collins, W., and M. Petit. 1998. "Strategic Issues for National Policy Decisions in Managing Genetic Resources." World Bank ESDAR Special Report 4. Washington, D.C.

Damania, A. B. 1996. "Swaminathan Foundation Holds Technical Consultation to Develop Framework for Farmers' Rights in India." *Diversity* 12:6–8.

Deolalikar, A. B., and R. E. Evenson. 1990. "Private Inventive Activity in Indian Manufacturing: Its Extent and Determinants." In R. E. Evenson and G. Rains, eds., *Science and Technology: Lessons for Development Policy*. Boulder, Col.: Westview Press.

Doyle, J. J., and G. J. Persley, eds. 1996. *Enabling The Safe Use of Biotechnology: Principles and Practice*. Environmentally Sustainable Development Studies and Monograph Series 10. Washington, D.C.: World Bank.

Eberhart, S. A., H. L. Shands, W. Collins, and R. L. Lower, eds. 1998. *Intellectual Property Rights III—Global Genetic Resources: Access and Property Rights.* Madison, Wis.: Crop Science Society of America.

Evenson, R. E., D. D. Evenson, and J. D. Putnam. 1987. "Private Sector Agricultural Invention in Developing Countries." In V. W. Ruttan and C. Pray, eds., *Policy for Agricultural Research.* Boulder, Col.: Westview Press.

Flack-Zapeda, J., and G. Traxler. 1998. "Rent Creation and Distribution from Transgenic Cotton in the U.S." Paper presented at the symposium, Intellectual Property Rights and Agricultural Research Impacts. CIMMYT, El Batan, Mexico.

Fowler, C. 1998. "Rights and Responsibilities: Linking Conservation, Utilization, and Sharing of Benefits of Plant Genetics." In Eberhart and others, eds., *IPR III—Global Genetic Resources: Access and Property Rights.*

Garcia, J. C. 1998. "Brazil." In M. L. Morris, ed., *Maize Seed Industries in Developing Countries.* London: Lynne Rienner Publishers, Inc.

Ginarte, J. C., and W. G. Park. 1997. "Determinants of Patent Rights: A Cross-National Study." *Research Policy* 26:283–310.

Gould, D. M., and W. C. Gruben. 1996. "The Role of Intellectual Property Rights in Economic Growth." *Journal of Development Economics* 48(2):323–50.

Grief, S. 1987. "Patents and Economic Growth." *Int. Rev. Ind. Property and Copyright Law* 18:191–213.

Hawtin, G., and T. Reeves. 1998. "Intellectual Property Rights and Access to Genetic Resources in the Consultative Group on International Agricultural Research." In Eberhart and others, eds., *IPR III—Global Genetic Resources: Access and Property Rights.*

Heinemann, A. 1996. "Antitrust Law of Intellectual Property in the TRIPs Agreement of the World Trade Organization." In Beier and Schricker, eds., *From GATT to TRIPs—An Agreement on Trade-Related Aspects of Intellectual Property Rights,* vol. 18.

Helpman, E. 1993. "Innovation, Imitation and Intellectual Property Rights." *Econometrica* 61: 1247–80.

Henderson, D. R., P. R. Voros, and J. Hirschberg. 1996. "Industrial Determinants of International Trade and Foreign Direct Investments by Food and Beverage Manufacturing Firms." In I. M. Sheldon and P. C. Abbott, eds., *Industrial Organization and Trade in Food Industries.* Boulder, Col.: Westview Press.

ISAAA. 1996. "Annual Report 1996: Advancing Altruism in Africa." International Service for the Acquisition of Agri-Biotech Applications, Ithaca, New York.

Jaffe, W., and J. van Wijk. 1995. *The Impact of Plant Breeders' Rights in Developing Countries.* Amsterdam: Univ. Amsterdam.

Kilman S., and S. Warren. 1998. "Old Rivals Fight for New Turf—Biotech Crops." *Wall Street Journal* (May 27):B1.

Knudson, M. K., and W. P. Hansen. 1991. "Intellectual Property Rights and the Private Seed Industry." USDA, Economic Research Service, Agricultural Economics Report 654.

Konan, D. E., S. J. La Croix, J. A. Roumasset, and J. Heinrich. 1995. "Intellectual Property Rights in the Asian-Pacific Region: Problems, Patterns, and Policy." *Asian-Pacific Economic Literature* 9(2): 13–33.

Lele, U. 1998. "Organizing International Science for Environmentally and Socially Sustainable Agricultural Development in a Globalized World: The U.S. Case." Paper presented at the 1998 American Association for the Advancement of Science 150th Anniversary Annual Meeting and Science Innovation Exposition, Philadelphia, Pa.

Leskien, D., and M. Flitner. 1997. "Intellectual Property Rights and Plant Genetic Resources: Options for a *sui generis* System." Rome: IPGRI, *Issues in Genetic Resources* 6.

Lesser W. 1987. "Anticipating UK Plant Variety Patents." *European Intellectual Property Review* 6:172–77.

————. 1994. "Valuation of Plant Variety Protection Certificates." *Review Agricultural Economics* 16: 231–38.

————. 1997. "Estimating Cost Components for Distributing Genetic Materials Held *ex situ.*" *Ass. Systematics Collections Newsletter* 25 (June):35–40.

————. 1998a. *Sustainable Use of Genetic Resources under the Convention on Biological Diversity: Exploring Access and Benefit Sharing Issues.* London: CAB International.

————. 1998b. "Comments on Transaction Costs, Trust, and Property Rights as Determinants of Organizational, Industrial, and Technological Change—A Case Study in the Life Sciences Section." Presented at Economic and Policy Implications of Structural Realignments in Food and Agricultural Markets: A Case Study Approach. Park City, Utah.

Lesser, W. H., and A. F. Krattiger. 1994. "Marketing 'Genetic Technologies' in South-North and South-South Exchanges: The Proposed Role of a New Facilitating Organization." In A. F. Krattiger, J. A. McNeely, W. H. Lesser, K. R. Miller, Y. St. Hill, and R. Senanayake, eds., *Widening Perspectives in Biodiversity.* Geneva and Gland: International Academy of the Environment and the World Conservation Union.

Lourie, L. S. 1998. "The U.S. Position on Developing Trade Agreements Concerning Intellectual Property." In Eberhart and others, eds., *IPR III—Global Genetic Resources: Access and Property Rights.*

Mansfield, E. 1995. *Intellectual Property Protection, Direct Investment, and Technology Transfer: Germany, Japan and the United Sates*. IFC Discussion Paper 27. Washington, D.C.

Maskus, K. E., and M. Penunbarti. 1995. "How Trade-related Are Intellectual Property Rights?" *Journal of International Economics* 39:227–48.

Mazzoleni, R., and R. R. Nelson. 1997. "The Benefits and Costs of Strong Patent Protection: A Contribution to the Current Debate." University of Vermont and Columbia University.

Meller, M. N. 1998. "Planning for a Global Patent System." *J. Patent and Trademark Office Society* 80 (June): 379–91.

Nogues, J. 1990. "Patents and Pharmaceutical Drugs: Understanding the Pressures on Developing Countries." World Bank WPS 502. Washington, D.C.

Ojwang, J. B. 1989. "A Regime for Protecting Inventions and Innovations." In C. Juma and J. B. Ojwang, eds., *Innovation and Sovereignty: The Patent Debate in African Development*. Nairobi: ACTS.

Patel, S. J. 1989. "Intellectual Property Rights in the Uruguay Round—A Disaster for the South?" *Economic and Political Weekly* (India) May 6: 978–93.

Perrings, C. 1995. "Economic Values of Biodiversity." In V. H. Heywood, ed., *Global Diversity Assessment*. Cambridge: Cambridge Univ. Press.

Posey, D. A., and G. Dutfield. 1996. *Beyond Intellectual Property—Toward Traditional Resource Rights for Indigenous Peoples and Local Communities*. Ottawa: IDRC.

Reichman, J. H. 1993. "The TRIPs Component of the GATT's Uruguay Round: Competitive Prospects for Intellectual Property Owners in an Integrated World Market." *Fordham Intellectual Property, Media, and Entertainment Law Journal* 4 (Summer): 171–266.

Reuters. 1998. "Zeneca Sues Monsanto." July 30. (Available at www.agriculture.com)

Roffe, P. 1974. "Abuses of Patent Monopoly: A Legal Appraisal." *World Development* 2:15–26.

Rusike, J. 1998. "Zimbabwe." In M. L. Morris, ed., *Maize Seed Industries in Developing Countries*. London: Lynne Rienner Publishers, Inc.

Scherer, F. M., and D. Ross. 1990. "*Indicators of Market Structure and Economic Performance*." Boston: Houghton Mifflin Co.

Schiff, E. 1991. "*Industrialization without National Patents*." Princeton, N.J.: Princeton Univ. Press.

Seiler, A. 1998. "*Sui generis* Systems: Obligations and Options for Developing Countries." *Biotechnology and Development Monitor* 34 (March): 3–5.

Sherwood, R. M. 1997. "Intellectual Property in the Western Hemisphere." *Inter-American Law Review* 28: 566–95.

Sherwood, R. M., V. Scartezini, and P. D. Siemsen. 1999. "Promotion of Inventiveness in Developing Countries through a More Advanced Patent Administration." *Patent World* (May).

Shiva, V. 1996. "Agricultural Biodiversity, Intellectual Property Rights and Farmers' Rights." *Economic and Political Weekly* (India), June 22.

Straus, J. 1996. "Implications of the TRIPs Agreement in the Field of Patent Law." In Beier and Schricker, eds., *From GATT to TRIPs—An Agreement on Trade-Related Aspects of Intellectual Property Rights*, vol. 18.

The Crucible Group. 1994. *People, Plants and Patents*. Ottawa: IDRC.

The Lancet. 1993. Editorial. 342 (8885) Dec. 11.

UNCTAD (United Nations Conference on Trade and Development). 1975. "The Role of the Patent System in the Transfer of Technology to Developing Countries." New York: UN, TD/B/C.6/146, Aug. 8.

Union of Concerned Scientists. 1997. "1997 Global Acreage of Monsanto's Transgenic Crops." *The Gene Exchange* Fall: 11.

Watal, J. 1996. "Introducing Product Patents in the Indian Pharmaceutical Sector—Implications for Prices and Welfare." *World Competition: Law and Economics Review* 20(2):5–21.

WIPO. 1990. "Exclusions from Patent Protection." HL/CM/INF/1 Rev., May.

————. 1998. "PCT Update" and "WIPO Update." *J. Patent and Trademark Society* 80: 269–87.

Woteki, C. E. 1998. "The U.S. Role in Food Security." In Eberhart and others, eds., *IPR III—Global Genetic Resources: Access and Property Rights*.

2. Perspectives from International Agricultural Research Centers

ISNAR

Joel I. Cohen, Cesar Falconi, and John Komen

Intellectual property rights (IPR) is a broad term for the various rights granted by law for the protection of economic investment in creative effort. The main categories of intellectual property relevant to agricultural research are patents, plant variety rights, and trademarks as well as trade secrets. In most developing countries policies for the application IPR to biotechnology products are still being formulated. Some countries have never explicitly excluded living material from patent protection. Others have recently adopted IPR for biotechnology, or are discussing IPR legislation in which the inclusion of living material is proposed. Currently, African countries are exploring the options and implications for agricultural research of national policy decisions on IPR and biotechnology.

One reason agricultural IPR attracts such debate is that agricultural development, including the release of improved planting materials, has benefited from a long history of public sector "public good" investment. At the core of this system has been the free availability of plant genetic resources. Increased IPR protection of agricultural research does not always seem consistent with either the long-standing tradition of public-sector investment or with innovations contributed by international agricultural research or by informal or indigenous communities. Many observers fear that invoking such protection destroys the "public good" nature of agriculture, especially as it relates to the needs of the rural poor.

Many developing countries have become members of the World Trade Organization (WTO), and are bound to introduce international minimum standards for IPR protection. The extent to which changes in the IPR legislation of developing countries will, in fact, lead to accelerated technology transfer and to greater domestic innovation in advanced technology remains to be seen. Consideration of stronger intellectual property protection will entail an analysis of costs and benefits—the costs involved being more foreseeable.

ISNAR (International Service for National Agricultural Research) began work on IPR with a research report (van Wijk, Cohen, and Komen 1993), and a series of policy dialogues structured to identify IPR-associated needs as perceived by delegations from developing countries (Cohen, Crespi, and Dhar 1998). Based on this information, IPR options were identified for research managers who must consider if and what protection is appropriate for a range of research-based innovations, whose needs such protection serves, and how to weigh expected costs and benefits. From these seminars, it became apparent there are at least four important positions that must be reconciled with regard to IPR:

1. Scientists' perceived needs to apply for IPR
2. Institutional goals and objectives
3. End-user interests regarding the innovation
4. National policy goals.

Seminar follow-up has included an expanded research agenda and an integrated approach to managing IPR for national agricultural research organizations (NAROs), focusing on institutional goals and objectives. This approach is implemented through management courses addressing IPR, in combination with other requirements facing agricultural research administrators in developing countries. The results of the ISNAR policy dialogue and survey process, which are used to make recommendations regarding the management of IPR in international agricultural research, are given below.

The ISNAR IPR Program

Beginning in 1994, the Intermediary Biotechnology Service (IBS) at ISNAR initiated a series of regional seminars to discuss options and implications of policy decisions regarding biotechnology. These seminars support the overall goal for IBS, which is to support innovation in agricultural research by responding to policy and management needs for biotechnology. The seminars are one means of encouraging the integration of biotechnology with broader national agricultural priorities. Policy seminars have been held in Southeast Asia (Singapore, September 1994); East and Southern Africa (South Africa, April 1995); West Asia and North Africa (Morocco, April 1996); and Latin America (Peru, October 1996). The specific objectives of the seminars were to:

1. Examine ways of integrating biotechnology into agricultural research, and the overall objectives for agriculture
2. Identify gaps and needs in research management and planning for agricultural biotechnology
3. Develop follow-up initiatives to address the findings and recommendations of the seminars
4. Identify potential regional actions, complementing the needs and capabilities identified by participating countries.

IPR was a prominent theme in the seminar discussions and was selected as a case-study topic for the session on "National Policies: Identifying the Issues" in Southeast Asia. This session started with a keynote presentation by an international expert outlining the main issues facing developing countries. The presentation was complemented by one providing the private sector perspective, and one giving the perspective of a regional NGO. National working groups then discussed the situation regarding IPR for their respective countries, and identified existing gaps and possible actions to address them. Examples of working group results, showing the wide range of concerns involved in policy decisions on IPR, are given in table 2.1.

Table 2.1   Southeast Asia policy seminar: conclusions from national working groups regarding IPR

Country	Identified gaps	Possible actions
Indonesia	IPR	• Elaborate patent law with guidelines for implementation • Draft IPR regulations
Philippines	IPR	• Update laws • Standardize guidelines (include bioprospecting) • Involve government organizations, NGOs, peoples organizations, in formulating policies/guidelines
	Inadequate information on available technologies and IPR	• Documentation (global) • Regional "IPR - gene banking"
Singapore	Plant Variety Rights	To be addressed at the regional level
	IPR	• Legislation forthcoming • TRIPs agreement
Thailand	IPR	• Develop Plant Variety Protection Act • Consider developing protection of traditional varieties and indigenous knowledge

Sources: Komen, Cohen, and Lee (1995); IBS Policy Dialogues.

*Proprietary Technologies, MTAs,
and Legal Environment*

New products derived through biotechnology, especially those coming from the private sector, are finding wide use in agriculture. This is made evident by the number of products reaching fields and markets in industrial countries, and the amount of ongoing cross-licensing of biotechnology products by commercial agricultural research organizations. Most of these inputs are protected through some form of IPR. However, development and use of protected materials is also occurring among public, national, and international agricultural research centers (IARCs) working for and with developing countries.

IARCs and the NAROs of developing countries are affected by these trends. Changes in proprietary rights as they relate to agricultural research are particularly relevant to international agricultural research, because of two related developments: the increasing importance of biotechnology in IARC and NARO research, and the growing position of the private sector in international agricultural research.

ISNAR was asked, on behalf of the CGIAR Expert Panel on Proprietary Science, to conduct a survey from December 1997 to February 1998 in which all responding Centers indicated current use of proprietary inputs for biotechnology research. Proprietary technologies and materials are those that are privately owned, managed, or protected through some form of IPR. In total 46

discrete technologies and materials were identified as in use within eight broad technology categories. Most Centers use these technologies and materials in research on several mandated commodities, with 166 applications of proprietary research inputs recorded. Of the technology categories surveyed, three had the broadest application across Centers: selectable marker genes, promoters, and transformation systems (table 2.2). Applications to cereals and noncereals are nearly uniform, whereas noncrop uses are more limited.

These data were reconfirmed with each Center prior to finalizing the report, and clearly demonstrate the important role proprietary technologies and materials have assumed in IARC research, as is true for advanced research centers globally. Although some proprietary technologies and materials are being applied in a research-only context, others become part of germplasm or products suitable for dissemination to NAROs, NGOs, and other IARC research partners, or even directly to farmers. The results of this survey highlight the challenging new environment within which the CGIAR centers operate, since many of the proprietary inputs have use restrictions. Restrictions on use may become evident at the research stage, or not until finished products are ready for dissemination.

For private-public partnerships to be effective, a clear understanding of the IPR implications of joint technology development and use will be

Table 2.2   Applications of proprietary technologies by responding IARC, grouped by commodity category

Technology category	No. of applications by commodity category		
	Cereals [a]	Noncereals [b]	Other [c]
Selectable markers	17	25	2
Promoters	18	14	3
Transformation systems	12	14	3
Insect-resistance genes	8	11	0
Disease-resistance genes	6	5	0
Genetic markers	4	4	2
Diagnostic probes	0	0	3
Others	7	6	2
Total	72	79	15

a. Maize, rice, wheat, sorghum, finger millet, pearl millet.
b. Bean, cassava, tropical forages, potato, sweet potato, chickpea, cowpea, pigeon pea, groundnut, lentils.
c. Diagnostics, livestock health, microbial systems, general technology development.
Source: ISNAR (1998); Related Survey.

Table 2.3   Examples of public-private sector partnerships in international agricultural research

International program	Private sector collaborator	Technology	Collaborating institute(s)
Agricultural Biotechnology for Sustainable Productivity (ABSP)	ICI Seeds (USA)	Maize transformation with *Bacillus thuringiensis* protein genes, for resistance to Asian stemborer	Central Research Institute for Food (CRIFC, Crops Indonesia)
	DNA Plant Technology (USA)	Bioreactor technology for micropropagation of banana, pineapple, coffee, and ornamental palms	• Agribiotecnologia de Costa Rica (ACR) • Fitotek Unggul (Indonesia)
Feathery Mottle Virus Resistant Sweet Potato for African Farmers	Monsanto (USA)	Transformation technology for the development of virus-resistant sweet potato	Kenya Agricultural Research Institute (KARI)
International Service for the Acquisition of Agribiotech Applications (ISAAA)	Monsanto (USA)	Transformation technology for the development of potato resistant to potato virus X and Y	Center for Advanced Research Studies (CINVESTAV, Mexico)
	Asgrow Seed (USA)	Coat-protein technology for the development of melons resistant to cucumber mosaic virus	• Research Center in Cell and Molecular Biology (CIBCM, Costa Rica) • CINVESTAV
	Pioneer Hi-Bred (USA)	ELISA kits for local maize viruses	National Research Center for Maize and Sorghum (CNPMS, Brazil)
ODA Plant Sciences Research Programme	Agricultural Genetics Company (UK)	Insect-resistance genes for potato and sweet potato	International Potato Center (CIP, Peru)

Source: Cohen and Komen (1995).

essential. Negotiations over rights regarding use should be addressed as soon as possible when embarking on research partnerships. Capacity for undertaking such negotiations presents a time-intensive investment, often requiring that scientists be taken away from their research because other in-house legal capacity has not yet been developed. A range of illustrative partnerships is presented in table 2.3.

The most common legal arrangement by which IARCs obtain or provide permission for use of proprietary inputs is through material transfer agreements (MTAs), followed by licensing. Staff at IARCs may wish to review and use standard formats for MTAs, given the range of legal obligations that are generated, the popularity of the MTA as a means of obtaining proprietary technologies, and the likelihood that MTAs will be formulated and imposed by the technology supplier.

When legal obligations are imposed by MTAs, a Center must be in a position to honor those obligations. For example, when confidentiality obligations are imposed, a Center must police the handling of the material supplied. This may require the establishment of a secure system of operation and place researchers and visitors under confidentiality obligations. In the event of a violation of the MTA the IARC is likely to be sued rather than the responsible individual, because the Center is the signatory to the MTA. Alternatively, the supplying firm may bar the violating Center from receiving further materials, which can damage the entire research program.

The increasing legal complexity involved in the supply of proprietary technologies may raise matters of contract law, intellectual property law, biodiversity and biosafety law, technology transfer, and competition law (if restrictive provisions are imposed in a licensing agreement). These legal complexities indicate the need for IARCs to have a primary legal administrator to ensure the compliance of staff with obligations generated by MTAs, as well as a Center's compliance with the terms of intellectual property licenses. Indeed for 40 percent of the 58 products identified in the survey, no information was available on possible use restrictions. That finding emphasizes the urgent need to manage better acquired technologies using knowledgeable legal counsel. Where several IARCs use the same category of proprietary tools, it may be advantageous to cooperate in the acquisition of those technologies. For example, the survey indicates that CaMV/35S is the promoter of first choice among the Centers, making opportunities for coordination evident.

New capacity is needed on the part of decisionmakers, managers, and scientists to provide clarity on biotechnology policy and research agenda. A series of management courses for research directors in selected Asian countries has been developed, building on past publications and the policy seminars that underscored the need for human resource development. The course ("Managing Biotechnology in a Time of Transition") is supported by the Japanese Ministry of Foreign Affairs. The first course was held in November 1997 in Indonesia, and included a session on "Managing International Technology Transfer and IPR." In the course we specifically concentrate on the practical management of intellectual property at NAROs, as reflected in the keynote paper prepared by Blakeney (1997). This portion of the course is integrated with negotiation exercises, and case studies illustrating IPR considerations for public-private partnerships, end-user considerations, and other management responsibilities.

Managing Institutional IPR

Selecting from among the several types of available IPR protection is a complex management decision. In many public organizations, offices of intellectual property have been set up to help with these decisions. One responsibility of such offices is to consider the accountability requirements and public expectations regarding innovations produced with public funds. These offices can also help the national research program to anticipate the need to scale up, develop, and move innovations into production (see Erbisch, this volume).

The principal objectives of an agricultural research institute's intellectual property management policy are to create a fruitful climate for innovation and invention, and to create a better understanding of the available legal rights for the protection of creative effort. As a general rule an institute will take the position that research results should be published and made available. However, the institute may be obliged to take steps to secure IPRs to its research output as a means of making it available to farmers. In some cases IPR considerations may delay research publication.

Given the complexities of the legal protection of IPRs and commercial arrangements for the acquisition and the exploitation of those rights, it seems appropriate to centralize aspects of the management of intellectual property. The research institute may choose between a specially dedicated research liaison office (RLO) or an official within the institute with intellectual property responsibilities. The decision will usually be one of cost and size. Where the research institute is associated with a university, the services of a central university institution may be available. For example, the University Business Centre of Universiti Putra Malaysia provides intellectual property coordination for the agricultural research institutes located on its campus. When not available, it will probably be advisable to deploy an officer to perform the intellectual property coordination function.

The main function of the RLO will be to determine if research results should be protected or not. In the cases that the office decides not to protect the innovation, the rights may be released to the innovator(s), usually subject to a nonexclusive license to use the intellectual property for research in the institute. In those cases where the office decides to protect the innovation, it will work with the innovator(s) in market evaluation and in finding commercial collaborators to exploit their intellectual property (table 2.4). The RLO will be responsible for securing professional assistance under appropriate conditions

Table 2.4   Patenting process and time line

Time line	Action by inventor (or RLO)	Action by patent office
Before first application	Invention and preliminary appraisal of patentability	
First application	• First patent application filed in home country • Establishing priority date	
Within 12 months after first application	• Further development of invention and technical/commercial assessment by internal staff, consultants, industrial, and government contacts • Decision taken to proceed or abandon, and costs estimated • Patenting route selected (national, European, international)	Official prior art search (novelty search)
12 months after first application	• Home filing consolidated • Foreign applications filed based on priority application	Official examination starts; precise moment depends on backlog
18 months after first application		Official publication of application (in some countries)
At a later moment variable in time	Further prosecution of patent application by applicant and attorney	Patent granted or refused

Source: Cohen, Crespi, and Dhar (1998).

of secrecy. The ultimate vehicle for commercialization may be the research institute, through the RLO, or through the institute's commercial arm. The RLO will ensure the coordination of commercialization activities and prevent conflicts of interest. In addition, the RLO would be also responsible for ensuring the compliance of a research institute with terms of any intellectual property licenses.

To enable the RLO to evaluate the commercial potential of inventions and discoveries, researchers should be obliged to report their research activities in a timely manner in an appropriate form. This form will capture all preliminary information required for initial evaluation, such as assessing the invention's patentability (or other intellectual property protection) and commercial potential, and fulfilling any obligations to research sponsors.

Due to the high cost of securing and then maintaining patents, the RLO must often delay filing of patent applications until strong commercial interest is evident. Applications may be filed in foreign countries depending on commercial potential and other needs.

Conclusions and Suggestions for Future Actions

Many IARCs and NAROs have formed partnerships to undertake joint or contractual biotechnology research, thereby receiving access to proprietary materials. These agreements reflect the changing nature of agricultural research. In such cases, the relevant technologies or licenses are held by partner organizations that may cover the use of proprietary inputs. However, such partnerships may only cover research. The use and dissemination of products may be a matter for further negotiation.

Experiences with accessed technologies make it clear that the international agricultural research system must adapt to the changing nature of agricultural research. With this evolution must come greater clarity regarding policies and practices in IPR. This is important for partnerships with the private sector, as well as for providing clarity to other stakeholders. Based on this need, and findings identified from the ISNAR policy seminars, recommendations are proposed for further consideration by the World Bank and for countries receiving loans. The Bank may wish to provide

- Support for a research liaison officer network, including capacity building, legal expertise, and consultations that review pros and cons of securing IPR protection for specific case examples
- Support for seminars/courses that stimulate learning and development of legal systems and expertise equal to that in industrial countries
- Advisory capacity representing the full extent of legal options (including MTAs), taking into account the public good nature of agricultural research, and where appropriate, issues regarding genetic resources
- Mechanisms that help develop and present clear policies and agenda for IPR and consolidate institutional knowledge for acquiring technologies.

References

Blakeney, M. 1997. "The Practical Management of Intellectual Property by Research Institutes." Paper presented at the ISNAR-IBS course, Managing Biotechnology in a Time of Transition. Yogyakarta, Indonesia.

Cohen, J. I., S. Crespi, and B. Dhar. 1998. "Should I Seek Legal Protection for My Research Results?" In S. R. Tabor, W. Janssen, and H. Bruneau, eds., *Financing Agricultural Research: A Sourcebook.* The Hague: International Service for National Agricultural Research.

Cohen, J. I., C. Falconi, and J. Komen. 1998. "Strategic Decisions for Agricultural Biotechnology: Synthesis of Four Policy Seminars." Briefing Paper 38. ISNAR, The Hague.

Cohen, J. I., and J. Komen. 1995. "Research Collaboration, Management and Technology Transfer: Meeting the Needs of Developing Countries." In D. W. Altman and K. N. Watanabe, eds., *Plant Biotechnology Transfer to Developing Countries.* Austin, Tex.: R. G. Landes Co.

Crespi, S. 1995. "Intellectual Property in Agricultural Biotechnology: Issues for Developing Countries." In J. Komen, J. I. Cohen, and S. K. Lee, eds., *Turning Priorities into Feasible Programs: Proceedings of a Regional Seminar on Planning, Priorities, and Policies for Agricultural Biotechnology in Southeast Asia.* The Hague/Singapore: Intermediary Biotechnology Service/Nanyang Technological University.

ISNAR. 1998. "Proprietary Biotechnology Research Inputs and International Agricultural Research." Report of an ISNAR Study Commissioned by the CGIAR Panel on Proprietary Science and Technology. Briefing Paper 39. ISNAR, The Hague.

Komen, J., J. I. Cohen, and S. K. Lee. 1995. *Turning Priorities into Feasible Programs: Proceedings of a Regional Seminar on Planning, Priorities and Policies for Agricultural Biotechnology in Southeast Asia.* The Hague/Singapore: Intermediary Biotechnology Service/Nanyang Technological University.

Van Wijk, J., J. I. Cohen, and J. Komen. 1993. *Intellectual Property Rights for Agricultural Biotechnology: Options and Implications for Developing Countries.* ISNAR Research Report 3. ISNAR, The Hague.

CIMMYT

Peter Ninnes

Rapid changes are taking place in intellectual property rights (IPR). At Centro Internacional de Mejoramiento de Maiz y Trigo (CIMMYT) we are endeavoring to stay abreast of developments so that our work for the resource-poor farmers of the world remains viable and relevant.

CIMMYT's primary objective in the past has been to develop germplasm. Our gene bank contains over 150,000 maize and wheat accessions, in addition to materials entrusted by developing countries over many years. During this period the global thinking on genetic resources has evolved, contributing to the increasingly complex and scrutinized environment in which we operate. For CIMMYT IPR issues relating to germplasm are critical and our responses to these issues have been positive and proactive. During 1997 we established an IPR Task Force to assist CIMMYT management in the following areas:

- Revision of CIMMYT's policy on intellectual property and material transfer agreements (MTAs)
- Revision of advice on the implementation policies
- Providing recommendations on potential agreements with the private sector.

The major change has been to distinguish clearly between "designated germplasm"— material designated as coming under the 1994 FAO–CIMMYT Agreement—and CIMMYT-developed germplasm, also referred to as CIMMYT "research products."

In May 1998 at the CGIAR Mid-Term Meeting (MTM) in Brasilia, an approach focusing on designated materials was endorsed. This approach replaces one previously adopted in 1995, an indication of the rapid changes in this area. Furthermore, steps have been initiated for full MTA harmonization among the CGIAR centers. However, IPR systems vary markedly from one country to the next, and private companies implement country-specific policies, implying that the Centers must allow flexibility in any harmonization policies.

CIMMYT and the Private Sector

The CGIAR centers' position statement on genetic resources, biotechnology, and IPRs developed at the 1998 MTM in Brasilia includes an agreed statement on principles involving Center interaction with the private sector and others on proprietary technology. Simply put, CIMMYT seeks opportunities to accelerate its work for the resource poor of the world, and this includes entering into partnerships with private companies (for example, see box 2.1 and appendix by Reeves).

CIMMYT seeks partnerships with the private sector that advance its mandate while providing opportunities for private companies to meet their own commercial objectives. In developing these partnerships, the CGIAR Guiding Principles and Center Policies must be translated into implementation strategies that facilitate the development and deployment of CIMMYT-generated technologies. This includes a consideration of nonexclusive licensing, although the benefits to private companies may not be immediately

---

Box 2.1   CIMMYT, apomixis, and the private sector

CIMMYT's apomixis research is a joint collaborative with ORSTOM. ORSTOM links French advanced research institutes and CIGAR centers with their developing country partners. The research program, aimed at the development of apomictic maize, involves one or more ORSTOM scientists being posted to CIMMYT, with additional staff and support provided by CIMMYT and ORSTOM. In 1996 a Mexican scientist joined the team through support from the Government of Mexico. The primary objective of the apomixis project is to transfer the gene(s) responsible for apomixis into maize from its closest apomic-

tic wild relative *Tripsacum*. ORSTOM filed for a patent in February 1997, an application co-owned by ORSTOM and CIMMYT (see appendix by Reeves), to protect this technology. The defensive patent was filed so that the technology can be managed for the benefit of resource-poor farmers through CIMMYT's partnerships with national programs and NGOs in developing countries.

Not surprisingly, the private sector has expressed an interest in the apomixis technology. CIMMYT has been engaged in extensive dialogue with companies interested in the technology and these discussions are continuing.

---

obvious. However, we believe there are areas where the Centers have considerable expertise and contacts that can prove beneficial, in particular to provide assistance in developing a regulatory framework for the deployment of new technologies.

Field and greenhouse trials with *Bacillus thuringiensis*/maize have been conducted in Mexico since 1995 and in the Philippines since late 1997. CIMMYT also works with the Kenyan Agricultural Research Institute (KARI) to develop a research program and protocol using transgenic maize in an integrated pest management (IPM) strategy to combat *Striga*, a parasitic weed. Negotiations are continuing with a company interested in evaluating its technology under field conditions in Kenya. To date CIMMYT has taken the view that up-front payments rather than future royalties are preferred; poverty alleviation and natural resource protection are the objectives, not the generation of income.

The "Defensive Protection" clause of the Guiding Principles for the CGIAR Centres on Intellectual Property and Genetic Resources states that:

The Centres do not see the protection of intellectual property as a mechanism for securing financial returns for their germplasm research activities and will not view potential returns as a source of operating funds. In the event that a Centre secures financial returns as a result of the commercialization by others of its protected property, appropriate means will be used for

furthering the mandate of the Centre and the objectives of the CGIAR.

CIMMYT currently uses 29 applications of proprietary technology in working with selectable markers, promoters, transformation systems, insect-resistance and disease-resistance genes, and genetic markers.

Lessons Learned, Lessons Offered

The following issues may be of interest to the World Bank as it forges new partnerships with its stakeholders, the CGIAR centers, and industrial countries:

1. Exclusivity—examples of mutual beneficial agreements with the private sector under nonexclusive arrangements should be highlighted as models for future collaborations
2. Legal—the establishment or support by the World Bank of a service facility or clearinghouse on legal and contractual matters would be welcomed, in the context of development assistance and loans. In particular this "honest broker" service should serve to create more confidence for national programs as they enter into agreements with Centers and the private sector
3. Distribution—(a) Seed information—There may be opportunities for the development of small, national seed companies to be fostered by World Bank financing as a component of other economic reforms, in part to compensate for differing multinational firm policies in "marginal" markets; (b) Seed

technology—Seed technology involving insecticides, (for example, *Bacillus thuringiensis* for corn-borers) should always include an IPM component. The World Bank could participate in direct or indirect support of the implementation of responsible IPM education programs; (c) Biosafety—The lack of a regulatory framework impedes the safe introduction of new technologies. It may be expedient for biosafety regulations to be developed on a regional basis rather than on a country by country basis. Through the assistance of the World Bank, the CGIAR centers and national programs could be organized into regional forums. The private sector also can benefit through the development of regulatory frameworks that are often cited as the bottleneck in the biotechnology delivery system.

Finally, the development of human resource capacity is an integral component of any new technology introductions. The World Bank has a long tradition of capacity building.

Appendix. Apomixis Research: Biotechnology for the Resource-Poor—Ethical and Equity Considerations*

Apomixis refers to naturally occurring modes of asexual reproduction through seeds. Apomictic processes lead to the production of offspring that are exact genetic replicas of their mother plant. Therefore, in contrast to sexual plants, apomixis offers a unique potential to fix favorable alleles and allele combinations in all subsequent generations. This could impact directly the current efforts to develop high-yielding varieties in developing countries using hybrid crops. Indeed, apomixis might potentially revolutionize deployment strategies for hybrids in all major food crops by (a) limiting the need to purchase seeds yearly; (b) allowing a drastic reduction in seed production costs; and (c) allowing for faster and more locally targeted breeding programs.

CIMMYT and ORSTOM have collaborated on a research program aimed at the development of

apomictic maize for over seven years. The primary objective of the ORSTOM-CIMMYT apomixis project is to transfer the gene or genes responsible for apomixis into maize from its closest apomictic wild relative *Tripsacum*. The research approaches involve two different and complementary routes. One is to transfer the genes for apomixis into maize from *Tripsacum* through a conventional "wide-cross" breeding approach. Alternatively, molecular strategies have also been developed to identify and isolate the apomixis genes in *Tripsacum*, in order to make them available not only for maize, but potentially for a broader range of food crops. Research initiatives are also under way to determine the appropriate strategies for the deployment of apomictic crops. This includes the definition of novel breeding schemes for apomicts; the evaluation of the potential economic impact of apomixis; and the assessment of the biological risk associated with the release of apomictic cultivars (that is effects on biodiversity).

Progress has been regularly reviewed. Recent breakthroughs have significantly increased the chances of success, but ultimate success is still a number of years away.

Why Is Apomictic Maize Important?

This research is conducted within the CIMMYT-ORSTOM project "Equity in Access to Hybrid Vigor for Resource-Poor Farmers." Because apomictic maize would not change genetically from generation to generation (unlike normal sexual hybrids or open-pollinated varieties), a single purchase of hybrid maize seed (with apomixis) could enable the resource-poor farmer to gain from the benefits of hybrid vigor in his or her crop, and then use the seed from the harvested crop to plant year after year. A once-only purchase could therefore confer significant yield benefits for 3, 4, 5 years or more. This is desirable from an equity viewpoint—unlike richer farmers, resource-poor farmers often cannot afford to buy hybrid seed each year. Harnessing new technology and focusing it on the needs of the poor also meets the high ethical standards of

---

* Presented by T. G. Reeves at a Workshop on Ethics and Equity in CGIAR's Use of Genetic Resources for Sustainable Food Security, Brazil, April 1997.

CIMMYT, ORSTOM, and the CGIAR. In a recent statement by Dr. Miguel Altieri, Chairman of the CGIAR NGO Committee, apomixis was cited as an example of the types of biotechnology most appropriate to the needs and circumstances of the resource-poor farmers.

Issues

To be effectively used for the benefit of resource-poor farmers, this technology should be freely available to them through the long-standing partnerships CIMMYT has with developing country NAROs and NGOs. This is the case with all technologies produced by CIMMYT.

However, ORSTOM and CIMMYT saw potential threats to this future free availability, given the likely value of this technology to commercial seed companies and the growing number of recent patent applications for related and unrelated inventions being filed, particularly by larger companies.

The traditional, and still preferred, approach to avoiding third parties patenting and "tying up" technology is public disclosure. This has often been used by public and not-for-profit organizations (for example, CIMMYT and ORSTOM) throughout the world. It is increasingly evident, however, that public disclosure is becoming less effective because broad patent applications are being used to protect wide domains of intellectual property around publicly disclosed inventions.

Public disclosure of key facets of the CIMMYT-ORSTOM work was imminent in February 1997 with the "defense" of a doctoral thesis based on this research. Given the overriding objective of ensuring that this technology, if successfully obtained, should be freely available for its intended beneficiaries—the resource-poor of the world—and in light of aggressive IP protection by others, ORSTOM filed a patent application in Paris on February 17, 1997. Any patent arising from this application would be co-owned by ORSTOM and CIMMYT.

CGIAR Guiding Principles

CGIAR has a set of working principles relating to intellectual property and genetic resources. Whilst these have not been fully endorsed by

CGIAR members, they have been formally noted as the framework within which the Centers carry out their work as custodian trustees of global genetic resource collections.

Under the CGIAR guidelines, there are specific references to patenting (points 10 and 11) that are pertinent to the apomixis filing. Immediately CIMMYT knew of the defensive patent application, contact was made with the CEO of INIFAP, the national agricultural research organization in Mexico, for two ethical reasons: (a) Mexico was in the process of becoming a participating partner in the apomixis research; and (b) Mexico is a long-standing and gracious host of CIMMYT's headquarters. A meeting was held between CIMMYT, ORSTOM, and INIFAP in March 1997, and there was unanimous agreement that the objective of all parties was to ensure that, if achieved, apomictic maize would be freely available for the resource-poor farmers of the developing world. A "hands-on" approach to proactively protect, for the poor, was seen as more ethical and equitable than a "hands-off" approach of doing nothing and hoping for the best (but expecting the worst).

This action has been communicated to colleagues in NARSs, NGOs, and advanced research institutions, and has been praised as an appropriate, responsible, ethical, and equitable approach.

Where to from Here? Some Dilemmas

At the least CIMMYT and ORSTOM have bought some thinking time for decisionmaking, but given the current IPR environment, not a lot! The patent application could be allowed to lapse at any time—the simplest, but not necessarily the most responsible, approach. However, to effectively implement the patent will be an expensive process, requiring funds that neither CIMMYT nor ORSTOM have, nor would want to divert from other research activities. An option being investigated is to contract with a third party interested in this technology, seeking a "win-win" outcome.

Early and preliminary discussions with the private sector have identified some interesting possibilities for such an outcome. For example, one scenario is as follows:
• CIMMYT/ORSTOM retain the IPR of apomictic maize

- Company A takes financial responsibility for execution and defense of the patent
- Company A provides "in-kind" additional resources (principally access to proprietary technology) to accelerate research and increase the likelihood of success of the apomixis program
- In return, CIMMYT/ORSTOM would license the technology for Company A to use solely in the industrial world. (Their use of apomixis would more likely be to increase the efficiency of plant breeding and seed production.) CIMMYT/ORSTOM would be able to (and indeed would!) make the technology freely available to partners in developing countries.

Such a scenario would be a "win" for resource-poor farmers, for Company A, and for CIMMYT/ORSTOM. It would also be an excellent example of a strategic partnership between public/private, and north/south, to harness technology in the fight for poverty alleviation, food security, and protection of our natural resources.

Conclusions

The environment in which CIMMYT and its partners work is changing rapidly. In order to remain effective and relevant and to meet our mandate it is necessary to respond to that change in many ways, including the establishment of effective partnerships with the private sector. CIMMYT is repositioning in such a way that high standards of ethics and equity remain paramount in our endeavors for the resource-poor of the world. We will make all future decisions accordingly.

3. Perspectives from Industry

Monsanto
Richard H. Shear

Biotechnology has been broadly defined as any technique or process using living organisms to make or alter products and/or improve plants or animals. This would include the production of microorganisms for specific uses, for example, the use of microbes to clean up oil spills. In its broadest definition it would include traditional plant breeding as well as molecular biology, including recombinant DNA. With the advent of recombinant DNA in the late 1970s and early 1980s, it was recognized that biotechnology held the promise of revolutionizing drug discovery, chemical manufacture, agriculture, and nutrition.

For the first time it was thought to be possible to make certain proteins having beneficially known properties that had been either impossible or too costly to make by traditional chemical techniques. Such compounds included human growth hormone, insulin, and the like. In agriculture companies such as Monsanto, Ciba-Geigy, Plant Genetic Systems, and others believed that crops could be "re-engineered" to provide technology for ensuring that the world could be fed as future population demands required. This belief was so prevalent that entrepreneurial start-up companies, such as Agracetus and Calgene, entered the race to find ways to improve agriculture. Serious R&D, however, did not really begin in the United States until the U.S. Supreme Court decided that although patent protection could not be extended to naturally occurring living beings, such protection was available to living organisms that had been altered by human intervention. In other words the fact that the claimed invention

was a living organism did not preclude its patenting (*Diamond v Chakrabarty, 447 U.S. 303, 206 U.S.P.Q. 193 (1980)*).

Although biotechnology held great promise in the 1980s, results in the form of products came slowly. Indeed, most products in the pharmaceutical area were not registered until the late 1980s and early 1990s. Agricultural biotechnology products in the form of genetically modified seed did not become commercially available until the mid 1990s. In both cases however, the products have been widely accepted and appear to be successes. Products such as erythropoietin, g-CSF, and human growth hormone are widely used. The products Bollgard® cotton and Roundup Ready® soybeans are in great demand by farmers. The speed at which these products have been discovered and brought to market, however, may be considered incredibly slow compared to the promise of the future.

Role of Genomics

In the future, biological advance promises to rival the advances made by the computer industry during the past 15 years. One of the reasons for this is the use of information technology to foster what is widely referred to as "genomics." Genomics is a term used for technological breakthroughs that involve the integration of basic and applied research in human and comparative gene mapping, molecular cloning, large-scale restriction mapping, and DNA sequencing and computational analysis. In simple terms genomics provides the ability to study the plant or animal genome, discover and map their genetic components, and annotate the genes with their function. This means we should be able to discover quickly which genes are responsible for which functions, potentially dramatically reducing the time required to determine bioactivity. Such breakthroughs are critical given the current timeline for conducting biological research of plants (9.5–18 years) and pharmaceuticals (5.5–11.5 years) (table 3.1).

With genomics the "bioprospecting" time can be cut dramatically. Although the timeline shows up to five years, it is of course understood that breakthroughs can occur quickly, slowly, or not at all. For every compound that is successful, 20,000 fail. It should be noted, however, that genomics will not shorten the remaining timeline beyond the bioprospecting stage. Clinical and safety testing will always be required, and the time required for such tests will not be reduced because of the great strides available via genomics.

While it is true genomics will decrease the amount of time for drug discovery, it is equally true that the investment is huge. Gene mapping, molecular cloning, large-scale restriction mapping, and DNA sequencing can be accomplished today, but not without investment in people, technology, machines, computers, and software. This will cost any company tens, if not hundreds, of

Table 3.1 Correct timetable for product development

Plant biotechnology timeline	
"Bioprospecting" starts	Day 1
Discovery of bioactive lead	1–5 years
Plant transformation	0.5–1 year
Greenhouse tests	1 year
Field tests	2–4 years
Toxicology/residue/safety tests	2–4 years
Backcrossing	3 years
Total	9.5–18 years
Pharmaceutical pipeline	
"Bioprospecting" starts	Day 1
Discovery of bioactive lead	1–5 years
Synthesis/formulation of bioactive	0.5 year
Initial studies	1 year
Clinical/safety studies	2–4 years
Registration	1 year
Total	5.5–11.5 years

millions of dollars. Although genomics holds great promise for drug and agricultural discovery, there is no guarantee that anything will be discovered or will be discovered before a competitor.

Furthermore, genomics does not lessen the risks or shorten the rest of the timeline. Neither does it reduce the other costs of bringing a pharmaceutical or agricultural product to the marketplace. Indeed, these costs continue to be several hundred million dollars.

Role of IPR

In order to provide the incentive for investment in such risky and costly R&D, intellectual property protection must be available. Without such protection who would undertake the risk and cost involved? The most important forms of intellectual property protection to encourage such research are patents and data protection.

Historically, patent and data protection was rather weak in many developing countries. In fact, some of the countries were havens for those who wanted to prosper from copying the products of inventors' hard work and good fortune. Indeed, the lack of compound protection for pharmaceuticals and agricultural chemicals in Brazil, India, South Korea, Taiwan, and others resulted in a great deal of legal counterfeit activity. As a result of the Uruguay Round of the GATT negotiations, and most notably the TRIPs Agreement, the environment is changing in many developing countries. Unfortunately, it is not changing rapidly, but change is coming.

Monsanto experienced problems in several countries during the 1980s regarding glyphosate, alachlor, and triallate production. Indeed, several situations occurred in which products being called AVADEX® and MACHETE® herbicides were sold by parties not licensed under any IPR. In one case the counterfeit product, that is nongenuine product using the proprietor's trademark, was diluted so that it did not have the correct amount of active ingredient, resulting in lack of weed control and yield loss.

If environmental, residue, toxicological, and other safety data are not kept secret and proprietary, others will copy the data and register to use it for their generic product. These data take many years and millions of dollars to generate.

Consequences of Weak IPR Protection

Generic production does stimulate local production, often at lower prices. Why, then, isn't it in the interest of a country to maintain a weak intellectual property system and encourage local manufacture of newly discovered products? First, most companies will not seek registration in a country which will not provide adequate patent protection or protection for efficacy, health, and safety data. Whatever the size of the local market, it cannot be large enough to compensate for the loss of data allowing competitors to access the marketplace in that and other countries.

Furthermore, the competitor did not spend the time and invest the money to conduct the research, and thus has an unfair competitive edge over the technology developer. While this may be considered a national benefit, it must be remembered that the new producer lacks health and safety data, does not have any experience in testing the pharmaceutical or seed product, and does not have the best formulation for the product. Any generic production will be at least 5–10 years behind the first introduction of the product in countries providing adequate intellectual property protection. Such delayed access has a definite, if hidden, cost.

If countries want the latest state-of-the-art technology, along with the assurance that the product works and is safe for its intended use, they must provide patent protection for all forms of inventions, regardless of how or where made. In other words patents should be available for chemical compounds, pharmaceuticals, genes, and animal and plant inventions. Further, countries must provide protection for registration data if timely access to state-of-the-art technology is expected. Needed protection includes limits on access to government files as well as controls over employee and former employee data leaks (trade secret protection).

Finally, any system must include reasonable enforcement mechanisms so that infringement can be halted. Without effective enforcement an intellectual property system will be used initially, but patent owners will discontinue introductions once they realize that proper protection is not really available.

Technology Licensing

The developer will sometimes introduce technology through local subsidiaries, but more often the technology is licensed to a local entity.

The important negotiated license terms usually include the scope of the technology to be licensed, the field of use or uses, transferability of the license or sub-rights granted by the agreement, compensation for the license, and the duties of the licensor and licensee, including how the technology transfer takes place. The licensing of biotechnological inventions presents some interesting issues when the product being licensed, genetically modified seeds, can self-replicate. In that case, control of the licensed product becomes of paramount importance. For example, can the licensor restrict the saving of seed by the ultimate customer, the grower? Although plant variety protection laws in accordance with UPOV provide for a limited farmers' exemption to save seed, it is apparent that any business based on genetically modified seed cannot survive long if customers need only buy the product once. Accordingly, without some form of saved seed restriction, the huge costs involved with agricultural biotechnology may be difficult to justify, especially in the case of varietal, nonhybrid crops.

If the invention is a genetically modified crop, it is often necessary to backcross the desired trait into locally adapted germplasm. This can often be done under a research agreement with local seed companies. Technology developers will be reluctant to enter such research agreements, thereby blocking transfer, unless there is protection against the loss or unauthorized distribution of the transformed germplasm, and an intellectual property system to provide enforcement. The loss of such technology is too easy and the risks too great. It is critical, therefore, in licensing agreements everywhere that enforceable restrictions on transferability of the technology, especially if it is germplasm, be agreed and adhered to by the licensee.

With respect to the actual transfer, it usually makes sense for the licensor to provide the education and training needed to provide a smooth, fast transfer. In agricultural products, this division of responsibility is complicated by the fact that the licensor has the knowledge of the technology, but the licensee has the expertise on local agriculture. For that reason the transfer of technology must be supported by the licensor and the local licensee. This applies independent of the form of the transfer agreement, license agreement, joint venture, or other form. The licensor is needed to ensure that the transformation of the desired trait is successful, the registration data are supplied, and the licensee and growers understand the proper ways to use the technology. The licensee is needed to backcross the transformed line into local germplasm and to market/distribute the final product.

Compensation for the technology is always a sensitive subject, but generally speaking the more risk that the licensee shares with the licensor the lower the royalty rate. If a licensee wants to access all of the benefits of 10+ years of work, and the millions of dollars spent, that licensee will be asked to pay relatively high sums. If, however, the licensee is involved from the start and shares in the work, the investment cost, and the risk, the ultimate compensation requested may be quite small. Thus, if developing countries want to ensure access to state-of-the-art technology at relatively low royalty rates, they should encourage local industry to become more involved at an early stage, especially collaboration with firms in the industrial countries. Adequate intellectual property protection will be needed to encourage both parties to collaborate.

Any contribution of germplasm as specified by CBD occurs early in the research process (table 3.1). In most cases the monetary value of the contribution will be dwarfed by the millions spent subsequently, so that the royalties for germplasm providers will generally be small.

When competitive conditions allow, many companies have provided technology to developing countries royalty-free or with very little payment. This can be done when the use of the technology will not compete with, or otherwise interfere with, the main commercial markets. For example, in Costa Rica, Monsanto transferred technology for use with plantains. In that case plantains were used only for local consumption and did not compete with bananas. It was a win-win situation for the company and Costa Rica. Companies have also provided training to developing country researchers so that technology can be developed for local use. For example, Monsanto allowed a researcher from Kenya to train at its laboratories and bring back to Kenya technology for sweet potato that ensured improved production of that local staple.

A World Bank Role

What can the World Bank do to encourage the rapid development of state-of-the-art biotechnology in developing countries? I see several things the Bank is in a unique position to do. First, the Bank can use its good offices to encourage the development of adequate intellectual property laws. Adequate IPR protection will encourage technology providers to license or otherwise transfer greatly needed technology. It will also provide incentives for local companies, at least in the more developed of the developing countries, to start investing in R&D of their own, as well as taking part in collaborations with other research-based organizations at an early stage. In that way, the country will benefit from importation of technology and the development of its own technology. This will result in less dependence on foreign technology, better balance of payments, and more competition in the global marketplace.

Second, the Bank can provide the funds necessary for countries to invest in early collaborative work with biotechnology companies. As noted above, by collaborating early and providing risk capital, the royalties for resulting technologies will be much lower.

Finally, the Bank can aid in training so that any technology transferred is properly used. Without sufficient training the benefits of any technology, including biotechnology, might not be fully realized. Indeed, without proper training in insect resistant management, for example, a valuable tool for insect control minimizing pesticide sprays may be lost through resistance development.

Monsanto, like many companies, looks forward to investing in the development of agriculture throughout the world. When intellectual property laws are improved, like those in Brazil, companies such as Monsanto will invest in biotechnology, manufacturing, and technology transfer.

AgrEvo
David L. Richer and Elke Simon

Poverty and hunger are not inextricably connected. Although poor, a subsistence farmer with sufficient land, stock, and seed will be able to feed his family. What the farmer cannot do, however, is feed the increasing number of landless neighbors. World population is expected to grow from 5.8 billion to about 8.5 billion by the year 2025. How will this vast population be fed?

The adoption of advanced agricultural technology in the industrial world has enabled farmers there to achieve considerable increases in crop yields without expanding areas under cultivation. Indeed, there is little additional cultivable land available. By contrast, increases in food production in developing countries have often been achieved, if at all, with limited benefit from scientific advances, and often by encroaching upon dwindling precious natural resources. A policy of "slash and burn" may provide more land for agriculture, but the benefits are temporary and the environmental costs no longer acceptable.

In the industrial countries we have technology that can help alleviate the food problems of developing countries. At the present time that technology lies mainly in the area of classical crop protection, but biotechnology will become increasingly important over the next 10 years. However, biotechnology alone will not provide a complete solution to world food shortages, and it is anticipated that chemicals will remain important in protecting crops from pests and diseases for decades to come. In fact it is likely that a combination of classical crop protection and biotechnology will provide the maximum opportunity for increasing yields and controlling pests and diseases.

In the area of classical crop protection industry recognizes the benefits of making its technology available to developing countries, but does so with some suspicion. Without strong intellectual property rights (IPR) including effective enforcement of regulations, licensing technology in some developing countries is tantamount to giving that technology away.

The developing countries, too, are suspicious. Although they are eager to receive technology, they fear the technology provider's demands for adequate IPR to protect that technology. There is a prejudice against IPR; it is widely believed that stronger IPR will increase prices, drain currency reserves, and place the farmer—and the nation—under the control of the multinationals. There is little, if any, basis for these fears, since the perception obscures the underlying truth that no farmer will pay more for a technology than the benefit it returns. In truth the technology provider and the farmer will benefit only by transferring technology at an acceptable price.

The divide between those who would provide and those who would receive is not unbridgeable. Technology transfer can be made simpler and more effective, but it will require greater understanding on both sides. As always the public sector will be necessary to assist in the bridge building.

Key Intellectual Property Rights

The problems created by negative perceptions of IPR (particularly patents and data protection) in developing countries are discussed below, as are the special problems that are presented by biotechnology. We do not pretend to have solutions to all of these problems. We do suggest areas where we believe that action is advisable—if not essential—and where funding by donor agencies will promote and assist the introduction of the latest technologies into developing countries.

Patents

A patent is perhaps the most important IPR involved in technology transfer. A license to make, sell, or use an invention protected by a patent is the simplest and most widely used method of transferring technology from one party to another.

Given the importance of patents in providing access to the latest technology, it is perhaps surprising that the patent system should be so poorly perceived in developing countries. One has only to look at the outcry in India and elsewhere when the American company, W. R. Grace, sought to obtain a patent relating to the extraction of an active component of the neem tree. Indian farmers had used the leaves of the neem tree for pest control for centuries, but alarmists—and others with their own agenda—created panic by suggesting that Grace would stop farmers from using their old remedies. This was arrant nonsense; no patent can stop a person from continuing something he has done before, but that fact did not halt the attacks against one multinational and the entire patent system.

It is against this background that the industrial world has to persuade developing countries to strengthen their patent laws. Progress has been made in the past three years, largely as a result of the Agreement on Trade-Related Aspects of Intellectual Property Rights (TRIPs); one of a bundle of agreements signed in 1994 under the Uruguay Round of GATT. TRIPs laid down certain minimum requirements for the protection of patents, and these minima are at last being included, although often reluctantly and with delay, into the laws of many developing countries.

TRIPs is a significant and welcome step toward strengthening patent protection. Patent laws must be further strengthened to permit claims to products, particularly compounds, and to plants, and also to ensure that patent holders can enforce their patents swiftly and with certainty. Procedural delays and unsympathetic courts often enable infringers to continue their abuse of the patented technology. These actions discourage further transfer of essential technology.

Those countries that have improved their patent laws have already seen an increasing number of collaborative efforts between western multinationals and local companies. There have been, for example, an increasing number of European companies entering into license agreements with Chinese companies for the manufacture of agrochemicals. Why should an improvement in the Chinese patent law make such a dramatic change? After all, China always was a large market, whether or not it had a strong patent law.

The cost of developing a new plant protection chemical is over US$150 million; the costs of developing a new transgenic plant are comparable. And the investment does not end there; product stewardship is required. Companies need to collect reports of adverse effects, and to train and advise local farmers, particularly in developing countries, to ensure products are used properly and safely. Product stewardship is expensive and labor-intensive, and the companies that sponsor their products responsibly in this manner must recover costs in the relatively short time, perhaps 10 years, of patent life remaining after the first marketing of the product (usually in an industrial country). It should come as no surprise, therefore, that companies are reluctant to enter those markets that provide little or no protection for their products, and thus allow generic manufacturers to benefit unfairly from the investment in research and the product stewardship. This was the situation in China under the former law. Companies were reluctant to permit local manufacture without any realistic means of control, and as a consequence, China was deprived of the technology. Now that China has brought its patent law closer to the European model and recognized the need for patent holders to be able to enforce their patents, companies from the industrial world are not only prepared to consider collaborative ventures with Chinese companies, but are actively seeking them.

There is considerable scope for the introduction and development of new products in developing countries. There are also many local needs and problems, often relating to crops. All of these require research, and often in research stations and universities. Research requires investments that companies are prepared to make only where they can see a benefit. Companies are far more likely to invest in those countries providing strong IPR where the results of their investment can be protected.

Of course, not all technology transfer projects are linked to commercial return. In agriculture there will always be a need for projects to improve the long-term position of poor farmers in developing countries. In this area the support and investment of the major funding agencies, such as the World Bank and FAO, play a major role. International agricultural research centers (IARCs) are a particularly efficient conduit for technology transfer. The achievements of all of these public bodies can be extended and accelerated by encouraging the private sector to match their investment and to play a role in the projects. Industry will, however, need some certainty—in the form of strengthened IPR—that its funding of projects will not undermine the investment that has already been made elsewhere.

Proprietary Registration Data

Crop protection products are subject to marketing and use approvals by national regulatory authorities. Before a product can be commercialized, companies must submit health, safety, and environmental data to the registration authorities who will decide on the suitability of the product for registration and local sale. The development of these registration data usually takes between 5 and 10 years, and the cost of data production alone may be as much as US$100 million. This is a high-risk investment since successful registration and sale of a new product is not a certainty.

The data provided to regulatory authorities are a substantial asset, and must be protected against unfair use by competitors. For example, the FAO Guidelines for the Registration and Control of Pesticides (Rome 1988) addressed the issue as follows:

> All data submitted by a company in support of its request for registration of its product should be treated as proprietary, and should neither be divulged nor used to evaluate the petition submitted by another applicant, unless by agreement with the owner of the data or unless a period of proprietary rights to the data has expired. . . . Apart from the injustice of allowing competitors to benefit from the use of data to which they have no right, the consequences of such an action would be to discourage, because it is unrewarding, the research and development required for the production of new pesticides which are needed, for example, for the control of new or difficult pests or to overcome resistance.

Almost all OECD member countries have included data protection in their national law, generally providing for at least a 10-year protection period for registration data. Unfortunately, many developing countries provide no such protection, although they still require submission of the data.

TRIPs attempts to deal with the issue, but the final provision (Article 39.3) dealing with data protection became so compromised during its drafting that it has had minimal positive impact on the protection of valuable data. Although countries bound by TRIPs are required to protect data-holders from unfair commercial use of their data, these countries are exempted from the requirement to keep data confidential where disclosure is "necessary to protect the public." That exemption is being interpreted by some developing countries as permitting their regulatory agencies to use protected data when considering applications by other applicants for similar products. In these countries, third parties may access the data, and achieve rapid approval to sell their own products without having made any of the investments required of the originator.

Rapid approval of imitation products may appear to be to the benefit of the local farmer, but in fact, it often results in a flow of substandard products with inadequate instruction for their use. Thailand is one country without a proper registration procedure and without data protection. Its market is flooded with cheap, poor-quality, and potentially dangerous products, whose unregulated use is causing serious environmental damage. The situation has been recognized by the Thai Government, and in cooperation with the Thai Crop Protection Association, the Government is in the process of introducing legislation providing for the registration of products and the protection of registration data. This will stimulate the introduction of modern plant protection products in an orderly market.

From the farmer's point of view, a generic manufacturer offering cheap agrochemicals appears to have only benefits. Even problems created by poorly formulated products or their improper use can be referred back to the originator, rather than to the generic supplier who, in any event, has no organization to deal with such problems. The originator, because of self-imposed requirements of product stewardship, feels obliged to incur the cost and trouble of resolving the situation, another benefit for the farmer. Where then is the disadvantage?

Without adequate protection for their data, companies will either not market their latest products in certain countries, or will defer entry until the products have become generic. Where a company has no market, it is unlikely to establish or invest in local research stations and distribution systems, or to undertake research into the needs of local crops, terrain, and climate, or to enter into collaboration with local research institutes and universities. The result is that the farmer has no access to the latest technology, and scientific education and know-how is not passed to those working on the ground who would improve the farmer's situation if they could.

Biotechnology has as yet made little impact on developing countries. However, when it does, it is likely that the regulatory authorities will require data on any new technology before sales are permitted. If these data are not protected, the technology owner will be reluctant to bring the new technology to developing countries, where it has potential for considerable benefit.

Biotechnology

The application of biotechnology to agriculture has taken longer than anticipated. Companies that began research in the early 1980s are only now seeing the results of their research come to the market, such as the recent introduction of herbicide-tolerant corn, soybean, and canola, insect-resistant corn, potato, and cotton, and hybrid canola. This market will expand rapidly in the coming decade, and the technology will be incorporated into other crops, particularly cereals, rice, sugarbeet, and wheat.

Biotechnology holds great promise for increasing food production. However, the science is complex, as are the IPR issues. For example, a single transgenic insect-resistant corn plant may involve patents on the transgenic plant, the method of transforming the plant, the genes contained in the plant, the modification of those genes, promoters to initiate the action of the genes, and markers to ensure that the genes are in place. There may also be Plant Variety Rights. Several different companies often own these various IPRs, and unless these companies can reach agreement—and freedom to operate—any one of them can block the sale of the plant.

As companies seek to include more traits into each plant, gaining freedom to operate will become even more complicated. The technology-based companies have recognized the problem, and considerable activity has lead to the restructuring of the seed industry and the evolution of the crop protection industry to a crop production industry. In the short term these changes will not assist the developing countries, but they do emphasize the need for the creation of indigenous high-technology seed industries. This should be a long-term development aim for all developing countries, and an aim that can be assisted by the public sector.

Seed is the vehicle by which this new technology will be delivered. Seed development has long been a goal of the IARC's classical breeding programs, which can now be improved and expanded with the powerful new biotechnological tools—if these can be transferred to the IARCs through collaborative research with the private sector. Collaborative research agreements should provide the IARCs with the latest developments in biotechnology and with training of personnel, but cooperating companies will require any resulting intellectual property to be protected. Such collaborations have already proved to be effective. For example

1. AgrEvo has entered into a collaboration agreement with an Indian research institute to evaluate specific natural products and their use in agriculture. AgrEvo provides know-how and funding. The Indian institute has the right to commercialize the results of the collaboration in India, and AgrEvo has the right to commercialize them in the rest of the world on the basis of shared profit.

2. AgrEvo's affiliated company, Plant Genetic Systems (PGS), has collaborated with the International Rice Research Institute (IRRI) and the International Potato Center (CIP) in joint R&D projects funded, respectively, by the Rockefeller Foundation and the Belgian AID Agency. PGS provided the parties with access to its state-of-the-art technology, and the projects were designed to develop this technology and know-how to yield results that were of practical use to both parties. The commercial rights to these results in industrial countries were retained by PGS, and both parties shared the commercial rights for the developing countries.

The IARCs can make germplasm, which has been improved with biotechnology and developed as a result of such collaborations, available to the NAROs. Direct funding from the public sector will attract greater interest and involvement of the private sector in these cooperative ventures. The possibility of obtaining patents for the results of such projects through strengthened patent laws will be a further incentive.

It is necessary to add a word of caution. Even when biotechnology is successfully transferred to the developing countries, a further problem remains. Genetically transformed plants must be marketed and cultivated with care, and as with classical crop protection, biotechnology will need product stewardship, and particularly, local management and advice. Without orderly marketing and use, the biotechnological traits will be rapidly lost or weakened, populations of resistant pests and diseases will develop rapidly, and the benefits of the new technology will be lost.

Techniques are in place for managing resistance. In the case of insects, the use of refuges or the use of multiple *Bacillus thuringiensis* genes will be important tools in resistance management, as will the occasional use of complementary insec-

ticide sprays. However, these tools will not avoid the potentially serious problem created by farmers saving seed as they have always done in the past. The protective traits in saved seed will, without quality control and by crossing with non-genetic crops, become weakened and lost. The growth of resistant populations of pests and diseases will be encouraged. This is a problem that can be alleviated by incorporating the technology into hybrid seed, which has the added advantage of giving higher yield. However, hybrid seed must be purchased afresh each year, and in poorer areas, there will be considerable opposition, even if there is money available, to purchasing seed each year. There is a role here for the World Bank and the IARCs.

Much of the responsibility for maintaining the efficacy and purity of genetically altered plants, and providing the necessary product stewardship, will lie with the breeders and seed distributors. The technology providers will, of course, have a major responsibility, although a presently confused one, since a seed may contain technology from several different providers. Where the technology is protected by adequate IPR—patents or data protection—so that there is a possibility of recovering some contribution to the investment in their technology, the technology providers will have an incentive to provide product stewardship as they do now with crop protection compounds.

Brumby, Pritchard, and Persley (1990) have succinctly described the role of the IARCs and the World Bank in furthering the use of biotechnology:

Because the costs of research in biotechnology are considerable, many of the biological processes involved and the novel genetic material they have produced are protected by patent rights that essentially extend the form of commercial protection already well established for crop breeders via plant variety rights. Research groups in developing countries will need interaction and collaboration with the private companies that are financing much of the current R&D. This will involve emphasis on partnership agreements in which the advantages of collaboration are clearly understood by both parties to the agreements. Much more joint R&D

among universities, public research institutes, and the private companies of developing and industrial countries appears likely to provide a promising approach to getting biotechnology into member countries of the World.

The arguments for such joint R&D have recently been demonstrated in Colombia. It has been reported there that a network of agricultural research institutions has developed a genetic block against the spread of a new barley rust with the potential to cause serious yield loss. This block was achieved in five years. Traditional breeding or the use of farmer-saved seeds would have taken tens, perhaps hundreds, of years to achieve the same end.

As policy analysts have maintained, and as the experience of AgrEvo and others confirms, the transfer of biotechnology to developing countries can be optimized by private companies working directly with IARCs, and with local institutions. International funding agencies should give serious consideration to the direct financing of projects between the private sector and IARCs, universities, and the like to enable them to become a conduit through which the new technology can flow to the developing countries.

National governments must also take some responsibility and should be urged to assist in the process of introducing the new technologies. In addition to strengthening patent and data protection laws mentioned, governments could take measures to encourage the development of an indigenous seed industry and assist in the process whereby germplasm containing the new biotechnology will flow to the farmer. Easing ex-

cessive quarantine laws, import restrictions on processing and storage equipment, high tariffs, and discriminatory seed certification legislation all need to be addressed. But finally, and inevitably, the availability of long-term hard currency financing will be essential.

Conclusions

We have attempted to provide a basis for discussion rather than to draw simplistic conclusions. The subject is certainly complex, but it appears to us that the World Bank sponsoring collaboration between the public and private sectors will encourage the transfer of technology for the benefit of agriculture in developing countries. It will be further encouraged by the strengthening of IPRs, particularly by providing for the possibility of patents for products and plants, for the more certain enforcement of patents, and for adequate periods of protection for registration data. IPR further provides an incentive for proper stewardship, necessary to provide effective and safe products and, in the case of transgenic plants, maintain that efficacy across many generations.

With greater understanding of the issues and with the aid of the World Bank, the public and private sectors can help ensure that two ears of corn can grow where one grew before.

Reference

Brumby, P., A. Pritchard, and G. J. Persley. 1990. "Issues for the World Bank." In G. J. Persley, ed., *Agricultural Biotechnology: Opportunities for International Development*. Wallingford, Oxon, U.K.: CAB International.

4. Perspectives from National Systems and Universities

Brazil

Maria José Amstalden Sampaio

Technology is increasingly important to the maintenance of international competitiveness and to national prosperity. The same is true at the company level (Chinen 1997). Technology development has become an international business, so that no nation or company can realistically expect to be self sufficient in technology, including agricultural technology. These factors, in turn, have intensified the significance of technology transfer, and of intellectual property rights (IPR). In the past, doubts were expressed about the acquisition of technology by developing countries from more advanced nations. Now, however, it is clear that developing countries must acquire some of those technologies necessary for their development, recognizing that in-house R&D can be much slower and more costly. A particular technology can often be obsolete by the time it is ready for commercial application, particularly in fast-moving fields such as biotechnology.

The extent to which IPR can be beneficial to science and its development depends very much on the way it is interpreted and implemented within the context of national laws. In agriculture/agribusiness any acquisition, transfer, adaptation, or joint development is and will be more confused until the issues, rules, and interfaces become clearer to all parties involved. For developing countries IPR issues could become a frightening "bottleneck" on the way to development. All the steps of the biotechnology development process involve inventions where IPRs can be obtained in many countries. Because of the broad scope of claims in "umbrella" patents,

the result has been a complicated web of ownership rights.

Brazil recently found itself in the middle of these confusing but important changes. On the one hand the government had committed to a strengthening of national IPR law. On the other Brazil is a major agricultural nation that buys and sells technologies developed by the public sector. Internal and external economic changes forced rapid adjustments in long-standing policies and practices in agriculture. The status and effects of these changes are outlined below. The short and longer term needs to provide an efficient and equitable transformation to a system with enhanced property rights are reviewed, in the context of an agricultural system long accustomed to public property resources.

IPR Legislation in Brazil

Brazil has a long association with IPR. Brazil was a founding member of the Paris Convention in 1883. The first "Brazilian Law" related to intellectual property was approved in 1809. More recently, an Industrial Property Code, approved in 1971 and in use until 1996, restricted the recognition of rights for pharmaceutical and food products and, not surprisingly, for biotechnology-derived ones as well. In 1993 Brazil ratified the Convention on Biological Diversity (CBD).

Patents

Following the international negotiations leading to the signing of the WTO (GATT)/TRIPs Agreement in late 1994, the National Congress gave more attention to a proposed new law. During 1995 public hearings took place with good local press coverage. This lead to the new Industrial Property Code being approved in May 1996. Law No. 9.279, also known as the "Patent Law," became fully applicable in May 1997. The new law gives legal protection to inventions related to pharmaceuticals, food processes, and biotechnology (Barbosa 1998).

The law excludes plants and animals and natural biological processes, but allows for the patenting of transgenic microorganisms that meet the three patentability requirements of novelty, inventiveness, and industrial application. Transgenic microorganisms means those organisms, except parts or the whole of plants and animals, that express a given characteristic that would not occur under natural conditions, unless due to human interference in its genetic composition. There remain many important issues related to the interface of biotechnology and chemistry that require clarifying decisions by INPI, which is linked to the Ministry of Industry, Commerce and Tourism.

In April 1998 the Brazilian Government published Decree No. 2.553 establishing the maximum percentage of compensation a potential inventor can receive from a public sector employer. Every public institution must now provide up to 33 percent (1/3) of the earnings from the licensing of a patent to the inventor. Implementation should follow soon, and a positive effect can already be observed: the measure is prompting scientists to speculate about better ways to help take a new invention to market.

Plant Breeder Rights

Since the "Patent Law" excludes plants, Law No. 9.456 was approved in April 1997. It became fully active in December 1997, following publication of the regulating Decree No. 2.366 and establishment of the National Service for Cultivar Protection (SNPC), linked to the Ministry of Agriculture and Food Supply.

The Cultivar Protection Law was prepared in accordance with the UPOV-78 Act, with two major differences. It provides protection to essentially derived varieties in accordance with the UPOV-91 Act, and specifies the free exchange (but not the selling) of seeds among small farmers' communities involved in Government-supported programs. Both articles present a challenge to implement. Time and practice will demonstrate how best to manage "essential derivation," and the best way to control seed multiplication and exchange (Sampaio 1998). The UPOV Council officially approved both modifications (with a few others) in April 1998, and Brazil requires only the approval of the National Congress to become a member of the UPOV Convention. By late 1998, nine species could be protected: maize, sorghum, rice, bean, wheat, cotton, potato, soybean, and sugarcane.

Scientists and business representatives founded BRASPOV, the Brazilian Breeders Association, to help implement the Cultivar Protection Law and establish breeders' legal rights. Many groups are getting together to develop the appropriate descriptors for cultivars of different species such as eucalypts, pines, ornamentals, vegetables, and fruit trees. This will help the national service increase the list of protected species.

Other Legal Developments

Although not directly linked with IPR, but certainly of major importance to the legal framework encompassing biotechnology and agriculture/agribusiness, the National Congress approved a Biosafety Law (No. 8.974) in 1995. Congress is also discussing a proposal to regulate access to biological resources ("Biodiversity Law"), which will also take into consideration the fair and equitable sharing of benefits with holders of traditional knowledge. One of the articles of the proposal that relates to IPR says: "no patent protection will be awarded to processes or products originated from accessions taken from the National Genetic Patrimony in disregard of the Law." These and related developments are highlighted in box 4.1.

The application of agriculture-related IPR is new to Brazil, compared with more than 40 years of experience in industrial countries. A less than perfect implementation is to be expected in such a short period of time. Some of the problems are related to a lack of specialized human resources and related infrastructure.

Box 4.1 IPR and related legislation being discussed or in preparation

Law Project No. 306—Access to Biological Resources and Associated Traditional Knowledge—submitted to Congress in November 1995

Project for a New Seed Law—submitted to Congress in 1998

Law Project for the Access and Use of Human Genetic Resources—to be submitted to Congress before 2000

Law Project for the legal recognition of Traditional Knowledge Rights—to be submitted to Congress before 2000

An Experience to Share

During the past three years a group of scientists, policymakers, and more than 10 Ministry representatives (designated as "GIPI"—the Inter-Ministerial Group for Intellectual Property) has been meeting to prepare IP law projects and related policies.[1] Overlaps have been identified between legal aspects of the TRIPs Agreement (WTO), CBD and its national implementation phase in relation to patent protection, cultivar protection, biosafety and biotechnology, biological diversity, and indigenous knowledge access. The group has also been able to make a productive link between the legislative and the executive levels of decisionmaking in the country, which has allowed for a faster advance in the implementation of IPR-related policies. Because of in-house participation during the formulation of the policies, scientists and policymakers are finding it easier to reproduce the information in their own institutes. As items are discussed and approved, seminars and workshops and the general media are helping to disseminate the information.

Effects to Date

Excessive market protection was one of the key elements restricting pharmaceutical sector development during the 25-year absence of patent protection. It was also expected that national firms would build up internal capacity unfettered by

property rights. When the Patent Law is passed, private investment by pharmaceutical companies is expected to increase dramatically.

A similar trend has been observed in the Brazilian seed industry. Following approval of the Cultivar Protection Law and the new Patent Law, many of the national private breeding programs are being absorbed by the multinational companies. It seems only a matter of time before more investment by private industry will take most of the commodity breeding programs away from government-funded institutions. Effects on the country's agriculture productivity and competitiveness, and the maintenance of investment to produce cultivars adapted to different ecosystems, remain to be seen. Environmental impact must be carefully monitored as well.

One of the justifications given by multinational companies for the acquisition of breeding programs and related businesses is the need for vertical control of the production chain to secure the ownership of genes and processes. A parallel explanation could be the need to acquire control of selected tropical germplasm to permit the introduction of proprietary genes in a better-adapted genetic background.

Preliminary negotiations are taking place between private multinational companies and governmental agriculture-related institutions represented by the larger universities and by EMBRAPA. Partners have little experience in this new approach, and it will be some time before the market challenges can accommodate all interests. The national seed companies that have not yet been taken over by the multinationals (a trend that is rapidly changing the face of the seed market) feel they are going to lose ground, and that it is only going to get worse with the incoming new genes made available through biotechnology inputs. The multinational companies have a strong interest in the growing Brazilian market, as a window to Mercosul countries (economic bloc formed by Argentina, Brazil, Paraguay, and Uruguay), wanting to secure their positions in a fast and aggressive manner. Partnerships and joint ventures will have to be analyzed as new cultivars are launched and grown in the next season. Brazilian farmers are eager to have access to the newest technology available in order to remain competitive.

IPR at EMBRAPA

EMBRAPA's (Empresa Brasiliera de Pesquisas Agropecuarias) "Institutional Policy for the Management of Intellectual Property," published in 1996, is summarized as follows:

- EMBRAPA shall maximize IPR use through the transfer or licensing of its proprietary technology, processes, and products (cultivars, software, CD, books, periodicals) but without compromising its social mission
- EMBRAPA shall seek legal protection for the technologies, processes, and products derived from its research program, giving credit to inventors when they are its own employees
- EMBRAPA shall authorize the use of its protected assets through a royalty-free license only when its social commitments are in jeopardy, and then only on approval of the Intellectual Property Committee
- EMBRAPA Research Units shall not release a new cultivar or disclose any process or product without a decision by the Intellectual Property Committee regarding the potential, convenience, and opportunity for IPR protection.

A clause of EMBRAPA's Genetic Resources Policy, which is being prepared to complement the Intellectual Property Policy, clarifies that EMBRAPA will not claim ownership of basic germplasm accessions received from other countries and held in trust for conservation and research purposes.

To implement the IPR policy EMBRAPA created a special Intellectual Property Committee (CPIE). It meets twice yearly to deliberate internal policies and other IPR issues related to processes, products, and technologies coming out of the research pipeline. A coordinating secretariat maintains an electronic mail link with members. During the initial phase, CPIE has been preparing rules for the functioning of laboratories, deciding on the need for confidentiality in projects, grant applications, and with external personnel such as ad hoc consultants, grantees, undergraduate and graduate students who develop joint research projects, international consultants, and visitors.

New challenges include the urgent prioritization of EMBRAPA's large domestication and breeding programs (more than 80 species), the review of ongoing contracts with the private seed industry, and analyses of the licensing of biotechnology tools already in use by research teams. Most tools currently in use are controlled by improper material transfer agreements (MTAs), including those allowing only for research use, and which have no provisions for possible future commercial use (see examples in table 4.1). This work should be advanced through the help of "local" IPR Committees (CLPIs), created during 1997 by all of EMBRAPA's 36 Research Units. Because of the diverse nature of EMBRAPA's research program, however, it has been difficult to train personnel fast enough. Units are scattered throughout the country and the issues involved with IPR implementation are complex. In May 1998 EMBRAPA decided to create a centralized unit, similar to the intellectual property offices commonly found in American universities. By September 1998 the Secretariat for Intellectual Property was fully functional. It will serve as the technology acquisition/transfer, negotiating, and licensing structure for those processes and products that have any interface with IPRs owned by EMBRAPA or by third parties.

While all these changes are taking place, EMBRAPA has been presented with proposals from private multinational companies interested in introducing and expressing their proprietary genes in the "elite" commodity cultivars (soybean, cotton, corn, and bean) owned by EMBRAPA. Farmers, long accustomed to having access to the latest technology, have been pressing EMBRAPA to launch cultivars containing the new transgenic traits loudly announced abroad. In addition to the new intellectual property rules, EMBRAPA must also comply with biosafety requirements. To comply with the Biosafety Law those EMBRAPA Research Centers working in modern genetic manipulation (transgenics) must form a special Internal Biosafety Committee (CIBio) and apply for the Biosafety Quality Certificate (CQB). The latter is necessary to qualify the laboratories to receive internal and external financial support. By the end of 1998, 16 research centers had active CIBios.

The value of EMBRAPA's "dominating" cultivars in the seed market has made negotiating the use of a few herbicide genes with multinational

Table 4.1   Proprietary technologies being used by EMBRAPA, June 1998

Specific tool (examples)	Form of protection	Research permission	Commercial use; possible constraints
CaMV/35S	Patent*	License under negotiation	Expected to be different for differ-rent shares of the market
Gus marker gene	Patent*	License under negotiation	License under negotiation
B. thuringiensis genes	Patent	License under negotiation	To be negotiated
Coat protein gene for bean	Patent deposited	No written agreement	N.K. Not anticipated
Coat protein genes for papaya	Patent	License under negotiation	Not anticipated for poor farmers. To be negotiated for other markets
RFLP marker genes	Patent	MTA	Unclear. To be negotiated
Kanamycin marker	Patent*	N.K.	N.K.
Hygromycin marker	Patent*	License under negotiation	Not anticipated
Imidazole gene	Patent	Research contract	License under negotiation
Glyphosate gene	Patent*	Research contract	License under negotiation
Glyphosinate gene	Patent	Research contract	License under negotiation
Fungal resistance genes	N.K.	MTA	To be negotiated
Nematode resistance gene	N.K.	MTA under negotiation	To be negotiated
Agrobacterium transformation system	Patent*	N.K.	N.K. To be negotiated
Biolistic transformation system	Patent*	No written agreement	Expected for export crops. To be negotiated

N.K. — not known.
*Patenting situation in Brazil still not made clear by INPI. Pipeline under analysis.

companies relatively straightforward. Others related to pest resistance and nutritional quality are under discussion. Each contract, nonetheless, is taking many months to develop because of the situation. The building-up of mutual confidence is an important step that takes time and effort. Companies are preparing for commercial release which, in the case of transgenic plants, will start in 2000-01. It is impossible to predict how these biotechnology-driven traits will be managed by Brazilian farmers. One of the critical challenges for EMBRAPA will be establishing royalty shares and licensing agreements. The interfaces of the intellectual property laws, one regulating the patent of the gene and another regulating the use

of the protected cultivar, are raising questions among researchers and managers. Each company is reacting differently to different proprietary traits, with the soybean business serving as the first model for negotiation.

EMBRAPA, by creating an alternative transformation protocol for Dicotyledonae plants through its Genetic Resources and Biotech Research Unit (CENARGEN), produced a technology "bargaining" chip that has changed the course of many negotiations. As a result EMBRAPA and possibly other national programs should invest in the development of their own genes/processes portfolio to gain easier access to proprietary technology owned by third parties. Many companies have recently shown interest in joint searches for new genes/molecules/microorganisms to help control tropical diseases and pests. Negotiations are being delayed because of the lack of appropriate legislation.

Implementing the IPR Policy within EMBRAPA has been a major challenge. The new legal background requires a dramatic change in the management of the Corporation's human resources. There are new opportunities and chances to stimulate scientific production through the distribution of royalties derived from proprietary technology. The new legislation, however, interferes with researchers' deeply rooted behavioral values, such as their need to publish and make readily available all research results. Keeping visitors away from laboratories and caring about the confidentiality of some sets of data have caused tremendous changes in the daily routine of researchers. A few scientists have easily jumped on the new bandwagon, but for most it will be some time before they can fully adjust. A strong capacity-building program will help develop a better awareness among researchers of the requirements and benefits of the new IPR program.

EMBRAPA determined some years ago that advanced biotechnology, and the development of transgenic crops, microorganisms, and even domestic animals, would play a central role in its goal of providing for increased sustainability and competitiveness of agriculture/agribusiness in Brazil. This would, in turn, contribute to the country's development and poverty alleviation. The increasing use of proprietary technology in agricultural R&D requires careful case-by-case analysis.

Although responding to the new IPR scenario, EMBRAPA continues to study and produce nonproprietary technologies that will be transferred to farmers and other clients without constraint. In fact most of EMBRAPA's technologies fall into this category. Maintaining trade secrets in agriculture is inappropriate in most cases, because major research projects are not impacted by the IPR ruling (for example, soil conservation and management, planting and harvesting methods, animal feeding programs, integrated pest management, and participatory breeding of social crops, such as cassava).

IPR and the CGIAR

During the May 25-29, 1998 CGIAR Mid-Term Meeting (MTM'98) held in Brazil, the Consultative Group on International Agricultural Research issued the International Research Centers' Position on Genetic Resources, Biotechnology and Intellectual Property Rights. Under the Guiding Principles (MTM'98 internal document) for the "Intellectual Property Protection of Designated Germplasm and Center Research Products," the document mentions that *materials supplied by the IARCs, whether designated germplasm[2] or the products of the IARCs' breeding activities, may be used by recipients for breeding purposes without restriction. Recipients, including the private sector, may protect the products of such breeding through plant variety protection that is consistent with the provisions of UPOV or any other* sui generis *system, and that does not preclude others from using the original materials in their own breeding programs.* That statement is very important for a country such as Brazil, which has exchanged multiple germplasm accessions with the IARCs over the past 25 years. Brazil has not only based its intense breeding program on plant material originating from many different parts of the world, but, as well, has contributed a large number of accessions to the system.

A related matter is the Centers' position statement on patenting, which reads: *cells, organelles, genes or molecular constructs isolated from material distributed by IARCs may be protected by recipients only with the agreement of a given Center,* and that *the Center will only give its approval after consultation with the country of origin of the germplasm where this is known or can be readily identified.* The statement adds that the consultation would include consideration of an appropriate benefits sharing agreement, whether bilateral or multilateral. This new policy is necessary to bring the CGIAR centers in line with the Convention on Biological Diversity (CBD) which has been signed and ratified by most of the CGIAR members and partner countries. Nevertheless, its implementation may hinder the immediate investment in the development of genes that could be screened and identified in designated germplasm samples. A necessary clarification is the application of this rule to samples obtained before the CBD came into force, and to what extent it will interface with the ongoing discussions related to the FAO Undertaking.

Conclusions

The introduction of IPRs in the agricultural sector in Brazil is quite recent, therefore there is little awareness of IPR issues among researchers, administrators (managers), and national companies linked to agriculture/agribusiness, or in any other areas of science and technology and industry in Brazil. Multinational companies are feeling their way around to find out how they can operate under the new legal regime. The Brazilian public has little understanding of the importance of IPR and, as a consequence, is not yet organized to deal with it. According to Briggs (1998), the role of the public sector on the impact of intellectual property laws could be substantial, but government policies on these issues are still unclear, in industrial and in developing countries alike. Actions urgently needed in Brazil, many of which would benefit from the assistance of the World Bank, include (G. E. Brandão, National Council for Scientific and Technological Development (CNPq), pers. comm.) (box 4.2):

1. Development of national competence in IPR, involving people at all levels, through

the granting of funded projects that could bring worldwide experts to teach formal courses to speed up the introduction of the IPR system.

2. Dissemination of IPR concepts and procedures through workshops, short courses, and seminars in Brazil, and through internships in patenting offices that support such training. This expertise related to biological matter is not available in Brazil at present. Recent experience has shown that even the well-established private patenting offices located in Rio de Janeiro and in São Paulo are only beginning to look at the new legal framework.

3. Promotion of better understanding and efficient use of IPR as an important tool for technological development.

4. Encouragement of the use of information contained in patent documents as an instrument for technological forecasting.

5. Establishment of offices in research institutes and universities to deal with IPR and to promote innovation.

6. Modernization of the National Institute for Industrial Protection (INPI) and the National Service for Cultivar Protection (SNPC).

7. Modernization of IPR capacity in the judicial system through courses and seminars.

8. Establishment of IPR-related research capability in the country.

9. Development of a culture and incentive mechanisms (including the legal framework) for venture capital that would allow for the "incubation" of technologically based businesses.

As a short-term answer for immediate problems EMBRAPA would welcome the continuation of the studies being developed by ISNAR (Cohen and others 1998) and recently presented to the CGIAR's IPR Committee. Many genes and processes were identified as being in use by IARCs and in need of negotiation with owners of the technology to be used and distributed to NARS (Cohen, Falconi, and Komen, this volume). If the information acquired by the centralized Advisory Unit to be created by the CGIAR (as recommended during MTM'98) could also be made available to help the NARS, the implementation of issues related to IPR might be greatly

---

Box 4.2   Suggested role for the World Bank in support of IPR implementation

1. Organize and finance specific training workshops and expert panels with the participation of developing country researchers.
2. Support the participation of developing country researchers in WIPO and UPOV training courses.
3. Support in-service training in IPR offices in the USA and Europe.
4. Encourage the implementation of a specialized electronic data bank to facilitate access to IPR-related issues.
5. Finance IPR-related studies to clarify the use of protected genes and processes in relation to the TRIPs Agreement and WIPO/WTO rules establishing the "freedom to operate" territory of the most common assets used by developing countries.
6. Promote the preparation of model MTAs and other technology transfer contracts.

advanced. An expert consultation on matters of contract, intellectual property, and competition law could be held electronically. Agreements with on-line databases, which, while presently available through the Internet, are not affordable by most research institutes, could serve as a strong encouragement for the use of the intellectual property system and development of the country's competence. The discussion of new policies to be established with American universities, regarding the development and licensing of proprietary products and processes by Brazilian researchers supported with Brazilian grants, could also greatly stimulate the development of new ventures.

EMBRAPA also welcomes the possible development of a pilot effort to discuss public-private relations with regard to technology transfer and IPR issues, involving the World Bank, the private initiative (agricultural multinational companies), and some developing countries, that could be done on a matching grant basis.

Finally, as MTAs and other licensing agreements become a part of the everyday life in publicly funded agricultural research institutions, the development of formal standard formats will help in the harmonization of basic terms between different parties. This should include the private sector, helping to guarantee the continuation of R&D projects that depend on the use of proprietary technology. Some of the clauses that have been proposed for MTAs by governmental research institutes and private companies, for example, deprive the researcher of any use of the technology in question except for research purposes. Other clauses claim ownership of any process or product developed with the use of the technology. Such trends in science may accelerate dangerous precedents that can conceivably culminate in partitioning and protecting all of the isolated bits of "useful" biology (Lee 1998). A wise balance is needed.

Notes

1. GIPI—the Inter-Ministerial Group for Intellectual Property is coordinated by the Civil House of the Presidency of Brazil and includes representatives of the Ministry of Science and Technology, Ministry of Agriculture and Food Supply, Ministry of Justice, Ministry of Industry, Commerce and Tourism, Ministry of Navy Affairs, Ministry of Foreign Affairs, Ministry of Environment and Legal Amazon, Ministry of Health, Ministry of Institutional Reform, Ministry of Economy and their linked institutes such as EMBRAPA (Agriculture), FIOCRUZ (Health), FUNAI (Indigenous People-Justice), INPI (Industry and Commerce), and IBAMA (Environment).

2. "Designated Germplasm" is a list of germplasm accessions, to be reviewed every two years, included in the Agreement signed on October 26, 1994, between each of the international agricultural research centers of the CGIAR system with the United Nations Food and Agriculture Organization (FAO), which placed germplasm collections maintained by that Center under the auspices of FAO.

References

Barbosa, D. B. 1998. "Uma Introdução à Propriedade Intelectual," vol. II. Editora Lumen Juris.

Briggs, S. P. 1998. "Plant Genomics: More than Food for Thought." *Proc. Natl. Acad. Sci. USA* 95 (March): 1986–88.

Chinen, A. 1997. "Know-How e Propriedade Industrial." Editora Oliveira Mendes.

Cohen, J. I., C. Falconi, J. Komen, and M. Blakeney. 1998. "Proprietary Biotechnology Inputs and International Agricultural Research." Briefing Paper 39. ISNAR, The Hague.

Lee, M. 1998. "Genome Projects and Gene Pools: New Germplasm for Plant Breeding?" *Proc. Natl. Acad. Sci. USA*, 95 (March): 2001–04.

Sampaio, M. J. A. 1998. "Propriedade Intelectual de Plantas: a Nova Lei de Proteção de Cultivares e suas Decorrências Imediatas." BIOWORK, A. Borém, ed. Viçosa, MG, Brasil.

India

Jayashree Watal

Intellectual property rights (IPRs) relevant to agriculture are identified and the IPR laws covering those rights are described, including India's international obligations to revise its IPR laws.[1] Public debate in India on the controversial IPRs, the status of applicable legislation, and prescriptions for public policy on IPRs and agriculture in India are analyzed. Other dimensions of the current IPR situation in India are discussed in an appendix to this paper by S. M. Ilyas.

IPRs Relevant to Agriculture

Several IPRs are particularly relevant to agriculture. These are patents, plant breeders' rights, trademarks, geographical indications, and trade secrets.

Patents are probably the most important because they provide, wherever they are available, the strongest protection for patentable plants and animals and biotechnological processes for their production. Patents give the holder the right to prevent third parties from making, using, or selling the patented product or process. Patented inventions must, however, be publicly disclosed in the patent documents. These provide a source of technical information enabling researchers to develop improved products or services. Products must meet the criteria of patentability: (a) novelty—that which is not known in the prior art; (b) nonobviousness—that which involves an inventive step; and (c) usefulness—that which is industrially applicable. With some differences, the patent laws of all countries follow these criteria. However, not all countries allow the patenting of plants, animals, or even microorganisms or of biotechnological processes relating to their production.

Presently, biotechnology appears to hold the most potential for productivity-improving advances in agriculture. Biotechnology R&D is largely concentrated in the hands of large multinational enterprises in the United States, Europe, and Japan. It is in this particular field of technology that proprietary rights over knowledge are increasingly important. Patents in the United States are granted for animals and for human gene sequences, if the criteria for patentability are met. The case law in the United States developed rapidly since the early 1980s following the grant of a patent for a modified bacteria that ate oil spills. This gave rise to the patenting of naturally occurring microorganisms, if a new, inventive, and useful technical intervention was included. Another landmark case (1987) was the patent granted to the Harvard oncomouse, useful in research cancer. The European Union has been slower to allow the patenting of plants and animals due, in part, to opposition from environmental activists in the European Parliament. This impasse has now largely been resolved with the imminent finalization of the new Biotechnology Directive by the European Parliament, authorizing the grant of patents to plants and animals with limited exceptions.

Many countries have developed plant breeders' rights (PBR) to reward conventional plant breeding efforts. The scope of such *sui generis* protection is weaker than patent protection, because it does not include the right to exclude third parties from making or using the protected material. The right holders can only prevent third parties from selling or commercially exploiting the protected material. The criteria used to grant such protection are also weaker than those used to determine patentability: (a) distinctness—distinguishable from earlier known varieties; (b) uniformity—display of the same essential characteristics in every plant; and (c) stability—the retention of the essential characteristics on reproduction. The novelty and nonobviousness requirements for patentable inventions represent a higher standard. PBR provides protection from direct copying and encourages breeding efforts in the private sector. Historically, in most countries, plant breeding was conducted by the public sector or by international research institutions. The institution of PBRs is relatively recent and was meant to encourage research by private breeders.

Marks used in commerce can be applied to agricultural and industrial products and services. For instance, trademarks are used to market seeds or spraying services. The essential purpose of a trademark is to distinguish the goods and services of one enterprise from another, thus preventing deception of the consumer. Such protection prevents the wrongful use of commercial marks and is not limited in time, although registration may have to be renewed from time to time. Almost all countries protect trademarks.

Geographical indications, including appellations of origin, are marks associated with products originating in a country, region, or locality where the quality, reputation, or other characteristics of the product are essentially attributable to its geographical origin. Most known geographical indications are applied to agricultural products, as in the case of wines and spirits. Protection of such marks prevents third parties from passing off their products as those originating in the given region. Famous examples are "Champagne" for sparkling wine and "Roquefort" for cheese from areas of these names in France, or "Darjeeling" for tea from this district in India. It is not necessary for these indications to be geographical names, as in the case of "Feta" for cheese from Greece or "Basmati" for rice from India and Pakistan. Plant varieties developed with traditional knowledge and associated with a particular region can conceptually also be pro-

tected as geographical indications. The advantage in such protection is that it is not time-limited, unlike the case of plant patents or PBRs. Commercial benefits, however, can be derived only when the name becomes reasonably well known.

Trade secret protection can be used by the agricultural sector to protect hybrid plant varieties, for instance, in the form of pureline stocks and/or crossing used, thus allowing a certain degree of appropriability even in countries that do not recognize PBRs. Trade secrets can be protected against third party misappropriation through laws relating to unfair competition or to restrictive trade practices or to contract law. In the United States trade secrets are protected by State and not Federal laws. Protection of trade secrets is not limited in time but, unlike patents or PBRs, does not apply if the secret is discovered independently by a third party. The advantage at least to the owner is that, unlike patents, there is no obligation to disclose the inventive or creative ideas to society, nor are there any application formalities.

Some industrial countries protect test data submitted for obtaining marketing approval of agricultural chemicals (pharmaceuticals are also given such protection in these countries) from use by third parties for a limited period of time, generally 5 or 10 years. Such protection gives exclusive marketing rights to the originators as an incentive to recover the investment made in testing such agricultural chemicals. Although developing countries also require the submission of such test data, no exclusivity is conferred on the originator for any period of time.

International Intellectual Property Law

Until recently the treaties administered by the World Intellectual Property Organization (WIPO) constituted the bulk of the international law on intellectual property. The relevant treaties for IPRs related to agriculture are the Paris Convention on the Protection of Industrial Property (1883 as revised up to 1967), and related treaties that deal with areas such as patents, trademarks, appellations of origin, or unfair competition. The Paris Convention establishes certain minimum standards and procedures for the treatment of industrial property, the most important of which are: (a) national treatment—the same treatment for nationals and foreigners; and (b) right of priority—the according of a grace period in the filing of industrial property applications in member states. However, it leaves considerable freedom to individual members to tailor laws according to national developmental and technological requirements.

The Union Internationale pour la Protection des Obtentions Vegetales (UPOV) has a multilateral treaty for the protection of new plant varieties, which it administers in cooperation with WIPO. The UPOV Convention (1961, as revised up to 1991) facilitates a uniform formulation of the extent and scope of PBRs. The 1978 Act was in force until April 1998, when the 1991 Act entered into force. There are at present 38 members of UPOV. The 1991 Act substantially enlarges the scope of breeders' rights and allows for restrictions of farmers' privilege and researchers' exemptions. The latter allows the use of protected materials in a breeding program, whereas the farmers' privilege permits farmers to retain the harvest as a seed source, both without the consent of the PBR owner. The 1991 Act also provides for a longer term of protection, and expands the universe of species/genera of plants for which protection is available, although this can be introduced in a phased way. Very few developing countries have instituted plant variety protection, and fewer are members of UPOV (Argentina, Chile, Colombia, Ecuador, Mexico, Paraguay, South Africa, and Uruguay (Source: *Diversity* 13(2 and 3) 1997, 3)).

The most recent international standard on IPRs is the Agreement on Trade Related Aspects of Intellectual Property Rights (TRIPs) of the newly formed World Trade Organization (WTO). There are now 132 members in WTO, with 30 more, including China and Russia, seeking membership. Although TRIPs obliges the adherence to the substantive provisions of the Paris Convention, it goes further on several aspects of industrial property laws. TRIPs obliges nondiscriminatory treatment in terms of national treatment between nationals and others as well as most-favored nation treatment among nationals of all WTO members. TRIPs further obliges members to either provide protection for plant varieties either through patents or through an effective *sui generis* law or through any combination of the two. Al-

though TRIPs calls for the institution of an effective *sui generis* system of plant variety protection, there is no reference to UPOV or a call to adhere to any version of it, making it the only exceptional case in TRIPs where the current international treaty on the subject is not mentioned.

TRIPs also obliges the patenting of microorganisms and microbiological and nonbiological processes for the production of plants and animals. It presently allows, however, the exclusion from patents of plants and animals and essentially biological processes for their production. In addition, no definitions have been provided in the agreement for the criteria for patentability (novelty, nonobviousness, and industrial applicability). These would be subject to interpretation by national patent offices.

TRIPs calls for "strong" process patents, strong in the sense that the rights of the patentholder extend to the product made by the patented process, and where there is a provision for the reversal of the burden of proof in any infringement proceedings. It is yet unclear whether such an extension of rights would imply rights over the product, even where such products are explicitly excluded, as is the case for plants and animals. These provisions, including those for plant and animal patents, are to be reviewed in 1999 when it can be expected that pressure will be brought to delete the exclusion for plants and animals, but the outcome is uncertain at this time.

The TRIPs Agreement also mandates a minimum level of protection of commercial marks such as trademarks and geographical indications. Geographical indications used on wines and spirits are given an absolute level of protection where use, even without the likelihood of deception of consumers, is prohibited.

For the first time in international law trade secrets have also been accorded the status of IPRs. The TRIPs agreement goes beyond the provisions of the Paris Convention on unfair competition, explicitly introducing trade secret protection in international law, and considerably strengthening it by extending the liability to third parties that induced breach of a trade secret. Further under the TRIPs agreement, test data submitted for obtaining marketing approvals of new pharmaceutical and agricultural chemical products is protected against unfair commercial use. The provisions of this section lend themselves to various interpretations (Watal 1998).

Under the TRIPs Agreement the protection granted for IPRs can be tempered by appropriate provisions on compulsory licensing or of competition law, particularly those relating to practices or conditions of licensing of IPRs that have an adverse effect on trade or transfer and dissemination of technology (Watal 1998).

CBD, concluded at the Rio Earth Summit in 1992, is a recent international treaty relevant to a discussion of IPRs and agriculture. Much discussion surrounds the Article 16 requirements for compulsory access to, and transfer of, technologies relevant to conservation under "fair and most favourable terms." There is a proviso, however, that such access and transfer shall be consistent with the adequate and effective protection of IPRs, so there is no reason to imagine the forced transfer of technology on any but commercial terms. Even the provision to cooperate to ensure that IPRs are supportive of, and do not run counter to, the objectives of CBD is subject to international law, which now includes the TRIPs Agreement. The fair and equitable sharing of benefits from the commercial use of genetic/ biological resources, or traditional/indigenous knowledge, is likely to remain as a good intention until there are legal instruments for their implementation.

Additionally, there are as yet no generally accepted means to reward what are called community IPRs (CIRs), that is embodiments of indigenous or traditional knowledge passed down over generations (Gupta 1996).

India's International Obligations on IPRs

As a developing country, India has a transition period of five years, that is up to 2000, for most provisions of TRIPs. An important exception is the introduction of product patents in areas of technology not covered so far, for which modifications an additional five years is granted. For pharmaceuticals and agricultural chemicals product patent applications must be accepted from January 1, 1995, and exclusive marketing rights must be granted for a period of five years or until the product patent is granted or rejected, on the fulfillment of the required conditions. Process by-product patents with the reversal of burden

of proof would, however, have to be in place by the earlier date. Such patents are similar in effect to product patents, with the difference that patent owners are not protected from patentable methods for the production of the product.

The current Patents Act of 1970 does not provide for the patenting of plants, animals, or microorganisms, not by exclusion, but because the definition of an invention seems to exclude them. Even microbiological processes are excluded if they involve a method of agriculture or horticulture, because such methods are specifically excluded. However, such applications have sometimes been granted patents, at least since the mid 1980s as is evidenced by the process patent granted to Agracetus, a U.S. company, on genetically engineered cotton cells and lines. This patent was later revoked in the public interest by the Government of India (Rao 1997).

Under TRIPs, India must either introduce patents for new plant varieties or have an effective *sui generis* law to protect them by 2000. India must also have strong patents on microbiological and nonbiological processes for the production of plants and animals. India has until 2005 to introduce product patents on microorganisms.

India must also bring the protection of trademarks, geographical indications, and trade secrets up to TRIPs standards by 2000. The current law on trademarks, the Trade and Merchandise Marks Act 1958, and the current jurisprudence, particularly under the common law tort of passing-off, is, by and large, in line with TRIPs. Marginal amendments are required, however, as in the case of the registration of service marks and the recognition of well known marks.

For geographical indications, the Trade and Merchandise Marks Act 1958 allows for the registration of certification marks, certifying quality or origin of a product. Such certification marks can be registered by an organization not producing the particular product, as, for instance, any association of producers or traders. In addition geographic indications are protected under the common law tort of passing-off. India would need new legislation to provide the higher level of absolute protection to wines and spirits required under TRIPs. In doing so, other Indian products or those of interest to India's trading partners can also be granted this higher level of protection, perhaps on the basis of reciprocity.

Although trade secret protection is available under contract law, and also laws on restrictive trade practices, India may have to introduce the legal basis to extend such protection to cover third parties who induce a breach of trade secrets. India would also have to legislate to protect undisclosed test data submitted for obtaining marketing approvals for new agricultural chemicals.

The Indian government has decided to join the Paris Convention. India also proposes to introduce national legislation to implement CBD through the Biodiversity Act, under which the terms of access to *in situ* genetic and biological resources would be governed (Business Standard 1998).

Public Debate in India

There has been extensive public debate, of an intensely political nature, on certain legislative changes required to implement TRIPs, including those related to agriculture. The latter include the institution of plant breeders' rights, patents for biotechnological inventions, and geographical indications. In addition, there has been considerable controversy over the implementation of CBD to establish the so-called "farmers' rights," and the fair and equitable sharing of benefits from commercialization of biological/genetic resources and traditional knowledge and practices originating from India. This public debate has been characterized by some degree of confusion over these various issues. Guided by a strong but narrow base of NGO activists, political parties or at least some leading political personalities have taken entrenched positions, forcing policymakers to consult such activists before finalizing legislation on IPRs.

Responsibility for policy on IPRs related to agriculture is widely dispersed amongst different ministries of the Government of India. The Ministry of Agriculture is responsible for legislation on plant variety protection only. Industrial property, including patents and trademarks, is the responsibility of the Ministry of Industry, the Department of Industrial Development. The Ministry of Commerce has taken the initiative on the legislation on geographical indications, because their protection is important for exports. This Ministry is in overall charge of trade negotiations

and agreements, including TRIPs. Trade secrets fall under the Ministry of Law as well as the Ministry of Industry (Department of Company Affairs). The protection of test data for agricultural chemicals is under the purview of the Ministry of Agriculture. The Ministry of Environment and Forests (MOEF) deals with CBD and is responsible for implementing national legislation.

Initiatives for the introduction of PBRs are widely recognized as having been spurred by private seed companies in India in the late 1980s after the adoption of the New Seed Policy in 1988. With this policy the Government of India liberalized the import of seed for joint ventures, including hybrid seeds, for a number of important crops. An empirical study has shown that such liberalization, including the development of hybrids, has had a positive impact on private R&D in this sector (Pray and Kelley 1997). Buttel, Kenney, and Kloppenberg (1985) have, however, forecast that the increasingly proprietary nature of plant biotechnologies and the decreasing role of IARCs and NAROs will adversely affect the diffusion of such technologies in developing countries. The two aspects of incentives for generation and diffusion of IPRs are not irreconcilable.

In some circles in India the new policies were seen as a victory for multinational enterprises, in spite of the fact that there were certain conditions regarding the transfer of the parent lines and critical breeding materials to the Indian partner of the joint venture (Bhattacharjee 1988). In particular, the TRIPs negotiations, especially in biotechnology, were seen as an attempt by industrial country multinationals to privatize the genetic diversity of developing countries (Menon 1991; and Sahai 1992). There were vociferous protests by some NGO activists against India's manner of conducting trade negotiations. The TRIPs proposals were seen as patenting of life itself, raising ethical as well as socioeconomic questions (Sahai 1992).

An association of farmers in Karnataka attacked the U.S. multinational seed company, Cargill Seeds, in early 1993, protesting the entry of multinationals into the domestic seed industry. It was feared that the prices of seed would skyrocket and threaten the food security of the country. This incident and the subsequent farmers' rally on March 3, 1993, at Delhi marked the height of the protest against the plant variety clauses of the TRIPs Agreement (the so-called Dunkel Draft). The Bharatiya Kisan Union (an all-India farmers' organization) even drew a parallel between these clauses and the takeover of the country historically by the British East India Company. The case of the patent on products derived from the neem plant was used to demonstrate the theft of traditional knowledge by multinationals, and the disastrous consequences for Indian farmers who would not be able to use neem seeds in traditional ways. It is only much later that some of the myths on the neem-based patents of W. R. Grace were adequately clarified (Gupta and MacAllister 1996).

The Ministry of Commerce attempted repeatedly to clarify that India did not have to patent plants, and that an alternative *sui generis* system could be devised to protect national interests. This did not placate some NGOs and other activists, however, who suspected that the term "effective" would be strictly interpreted to require patent-like protection (Shiva 1993). Even an article written by the then Director General of GATT, clarifying that standards contained in UPOV 1978 which allowed the farmers' and the breeders' privilege could reasonably be said to constitute effective *sui generis* protection, failed to assuage these fears.

Concurrently, some scientists and farmer activists concluded that India was capable of turning the TRIPs proposals to its advantage due to available skilled personnel, variety of agroclimatic zones, and facilities in agricultural research. This group asserted that farmers had nothing to fear, and may only benefit from the implementation of these proposals.

The M. S. Swaminathan Research Foundation, Madras, and the Research Foundation for Science, Technology and Natural Resource Policy, New Delhi (later changed to RFSTE, for Research Foundation for Science, Technology and Ecology) were consulted by the Ministry of Agriculture on the 1993 draft legislation on the protection of plant varieties (Shiva 1996). The M. S. Swamina-than Research Foundation prepared alternative draft legislation relating to plant breeders' and farmers' rights, which was discussed at a workshop conducted by that organization in late 1993. This draft attempted to reconcile the TRIPs Agreement with CBD and FAO's International Undertaking on Plant Genetic Resources 1989. It called for the establishment of a National Community Gene Fund as a mechanism for implementing farmers' rights based on a seed surcharge. Through this approach, India hoped to show the way to an international agreement by attempting to first evolve this concept in national law (M. S. Swaminathan Research Foundation 1994). The RFSTE went further and suggested that farmers' rights should form a part of the PBR legislation and should limit the scope of PBRs generated by the seed industry.

Given the public outcry on plant variety protection, the Government of India opened the draft legislation for debate in early 1994. This draft was bitterly criticized as following UPOV 1978, although it did include provisions on community rights and farmers' rights and extensive provisions on compulsory licenses (Srinivas 1994; Sahai 1994).

The Ministry of Environment and Forests came under tremendous pressure from public action groups to institute implementing legislation for CBD. It proposed legislation on biodiversity to regulate access to *in situ* genetic and biological resources, on conditions of prior informed consent, on fair and equitable sharing of benefits, and on transfer of technology on fair terms. Many feel that traditional knowledge should be registered so that it is not incorporated into patents without the knowledge or consent of the concerned communities. Consent would be given only after ensuring fair and equitable sharing of benefits. Others view rural, contemporary innovations as important for progress in agriculture, and advocate the institution of a new system of IPRs (for example, some kind of a global registration system (Gupta 1996)). Following the experience on the legislation for plant variety protection, it was decided to constitute a committee headed by Dr. M. S. Swaminathan, to include all the major stakeholders, including scientists, NGOs, environmentalists, and other relevant government departments/ministries. Regional seminars were held to discuss the various issues involved in the legislation, although the draft bill itself has not been made public. The issue of community rights may be resolved now in the proposed Biodiversity Act and not in the legislation on plant variety protection, although there is still considerable confusion on this issue. This is, however, being strongly opposed by NGOs that have been active in this debate, such as the RFSTE and the Gene Campaign. The process of consultation is still on in the Ministry of Environment and Forests.

Revised plant variety protection legislation, removing farmers' rights but retaining clauses on farmers' privilege and breeders' exemption, was attempted in 1997. This revised draft has been criticized as being modeled on UPOV 1991 deleting the farmers' rights altogether (Dhar and Chaturvedi 1998). There seems to be either little awareness that the draft biodiversity legislation incorporates farmers' rights, or a conviction that farmers' rights have to necessarily be juxtaposed against the PBRs granted to seed companies within the same legislation. It is not yet clear how this issue will be resolved.

In the meantime the Economic Times (1998) reported that some major European plant breeders have threatened to deny access to new rose varieties to Indian floriculturists if there is no protection of breeders' rights. The breeders' concerns are not only because of the royalties lost, but also because the effect on the quality of the flower, if illegal propagation and multiplication of the variety was allowed.

Similar exercises to involve the stakeholders in the drafting of legislation on biotechnological inventions have not yet been initiated by the Department of Industrial Development. That Department is charged with the task of amending the Patents Act 1970 to bring it in line with TRIPs by 2000. The public debate on this subject has so far been inadequate for the preparation of draft legislation. This is also the case for the protection of undisclosed information, whether trade secrets or test data.

The issue of geographical indications became controversial in the wake of a 1997 patent granted in the United States to a U.S. firm, on the claim of novel basmati rice lines and grains. In 1996–97, India exported about 490,000 metric tons of basmati rice valued at about US$358 million, constituting over 60 percent of the value of India's total exports of rice. In this case most Indians believe that India should have a strong law to protect geographical indications so that Indian names are not patented and misused in other countries (box 4.3).

There is a widespread belief in India that unless there is domestic *sui generis* legislation to pro-

tect geographical indications, these marks cannot be protected in other countries. TRIPs does allow WTO members to deny protection to geographical indications that are not protected in the country of origin. India, however, does permit the protection of such marks through certification marks as well as under the common law tort of passing off, provided it can be proved that the consumer would be deceived. The problem is that the certification mark system, or even any *sui generis* legislation, requires the definition of the product being protected. The delay in according domestic recognition to the mark "basmati" is probably more the result of the rice producers of India being unable to come to an agreement on the definition of the mark "basmati," rather than because the government has not passed *sui generis* legislation. Furthermore, protection of "basmati" as a geographical indication even in the U.S. would not have prevented Ricetec from patenting the new formulation in the U.S., but only prevented from the use of the name "basmati." In any event, the government has under consideration draft legislation to protect geographic indications in order to meet its TRIPs obligations by 2000.

Conclusions

The classical IPRs relevant to agriculture are patents, particularly on biotechnological inventions, plant breeders' rights, trademarks, geographical indications, trade secrets, and the protection of undisclosed test data. Farmers' rights and community IPRs are the new forms of intellectual property at the stage of initial conceptualization, where international or national laws are yet to evolve. India is not a member of the Paris Convention or UPOV, but is a member of WTO and is therefore obliged to implement the TRIPs Agreement within the time limits set out therein. Most of the TRIPs obligations on these relevant IPRs, including strong process patents for biotechnological inventions, have to be in place by 2000, and it is only for product patents on microorganisms that India has the additional time up to 2005.

Although legislative exercises on a *sui generis* system of plant variety protection began in 1993, the legislation is yet to be finalized. More recently, India has proposed the enactment of a

---

Box 4.3   Patent on Basmati rice

In September 1997, Ricetec was granted a U.S. patent for claiming novel basmati lines and grains that were created from the crossing of the basmati germplasm (of Pakistani origin), taken from an *ex situ* gene bank in the U.S. with an American long-grained variety. Ricetec claimed that the new varieties have the same or better aroma, grain length, and other characteristics of the original basmati variety grown in India and Pakistan, but can be grown successfully in specified geographical areas in North America.

The patent grant came to the notice of the Government of India in February 1998, and an Inter-Ministerial Committee was set up under the Secretary, Department of Industrial Development. The Agricultural Export Development Agency (APEDA) of the Ministry of Commerce has been entrusted with the task of representing the rice exporters in any reexamination of the patent in the U.S. Patent and Trademarks Office (USPTO), if it is decided that there are sufficient grounds for the reevaluation of the patent grant. The Council for Scientific and Industrial Research (CSIR), which successfully opposed and obtained the revocation of a patent on turmeric in 1997 in the USPTO, is assisting as well.

In India the question is: can Ricetec (or any other company) use the name basmati to sell rice that does not originate from India or Pakistan, or can basmati be protected as a geographical indication? Ricetec has claimed that basmati is a generic name denoting a variety of rice which, if labeled or advertised as "Ameri-

can made basmati type rice" or "basmati style rice," with a clear indication that the product originates from the U.S., does not deceive the public. The TRIPs Agreement accords absolute protection against the use of geographical indications with the words "type," "style," "kind," only to wines and spirits and to no other commodity. In addition if the Courts in the U.S. rule that the name "basmati" is already generic, denoting a variety of rice not necessarily associated with any geographical region, there would be no protection available for it. This is not yet tested in the Courts in the U.S., although APEDA is opposing the registration of the trademark "Texmati" by Ricetec in the U.K. on the grounds that it would deceive the consumers as rice originating from India and Pakistan. The fact that GAFTA of U.K. strictly ensures that "basmati" can only be used for rice originating from India and Pakistan should help India's case. The case has not yet been decided in the U.K. Trademarks Registry.

Some observers have expressed an opinion that a patent derived from the basmati germplasm amounts to biopiracy by Ricetec. However, it must be noted that the germplasm was taken from an *ex situ* collection in the U.S. Under current international law there is no prohibition on the exchange or use of germplasm for commercial purposes or otherwise. Therefore, India cannot claim sovereignty over germplasm already outside its jurisdiction.

Source: Media reports, February–April 1998.

---

biodiversity law to implement CBD, which is in the process of being debated and finalized. An important question is whether farmers' rights and community rights will be included in the plant variety protection law or in the biodiversity law or both.

Since the Government of India, as evidenced from its policies since the mid 1980s, wants to encourage investment by private seed companies, PBRs would help by giving incentives for private research. The issue of whether public sector research institutions should be allowed proprietary rights over their research is still controversial, although having such rights and also disseminating these technologies at reasonable prices are not necessarily contradictory. At the ICRIER Seminar many participants agreed that such proprietary rights would enable public sector research institutions to pre-empt private sector seed companies from not sharing commercial

benefits on varieties derived from them, and also maintain advantages in cross-licensing the results of their research. Steps would have to be taken to ensure that reasonable compensation is allowed to plant breeders. The deployment of skillfully drafted provisions on compulsory licensing and government use, and the recognition of the mutual interdependence between public sector and private sector research efforts, may resolve the apparent contradiction between incentives for research and the subsequent diffusion of such technologies at equitable prices.

CGIAR centers already assist in the diffusion of technologies, while ensuring reasonable compensation to the rightful owners. Several modalities have already been envisaged, such as material transfer agreements (MTAs), licensing or cross-licensing, joint ventures, or private funding of basic research in the public sector (Lesser *et al.* this volume). IARCs can play a further constructive role in the two-way transfer of technologies between NAROs and private sector seed companies, through the establishment of technology rights banks where IPRs and other rights over important technologies are purchased outright by IARCs, and distributed through NAROs to farmers in developing countries.

On the issue of patents granted on the basis of traditional knowledge without acknowledgment, there appears to be no solution but to document all such knowledge. The National Bureau of Plant Genetic Resources has set up a base collection of 160,000 samples of germplasm of various crop species in a National Gene Bank. It is expected to become one of the largest *ex situ* collections in the world (National Academy of Agricultural Sciences 1998). The state government of Karnataka, in collaboration with the Indian Institute of Science, Bangalore, has also launched a plan to map the biodiversity and traditional knowledge in its jurisdiction. In addition, the CSIR in India has already begun with a program to systematically document at least 400 species of plants where therapeutic, agricultural, and other uses are known (Indian Express 1997).

India has suggested in the WTO Committee on Trade and Environment that, under TRIPs, there should be an obligation on patent applicants, of biotechnological inventions based on genetic/biological resources or on traditional/ indigenous knowledge, to disclose the country of origin and to reveal whether the applicant has prior informed consent (WTO 1996). This suggestion was also proposed for the European Parliament's Biotechnology Directive but was rejected by the Commission as going beyond its international obligations. Such a solution is necessary in international intellectual property law, if developing countries are to be notified and fairly and equitably compensated for resources and knowledge taken from them for commercial benefit. There is an urgent need to build international consensus on this issue.

The legislative exercises on amending the Patents Act 1970, particularly on the patenting of biotechnological inventions, should be made more transparent, with the involvement of all stakeholders (scientists, farmer groups, private sector seed companies, lawyers, and NGO activists). Similar exercises are required to implement the

TRIPs provisions on undisclosed information. This would not only require the conduct of workshops and the setting up of drafting committees, but also the building up of mutual trust and respect, without which these would remain empty exercises.

There has been a recent vocal demand made by sections of the media to introduce *sui generis* legislation for the domestic protection of geographical indications such as basmati rice. The fundamental issue, however, is protection for Indian marks with India's major trading partners, a possibility available through existing laws of those countries. In addition, a conscious effort needs to be made to invest in and build up the brand equity of Indian markets in order to ensure that such marks do not become generic. Further, India should seek to conclude bilateral agreements with interested WTO members within the framework of TRIPs, to give better protection to products of mutual interest on a reciprocal basis. As long as this is done for specific geographic indications, and as long as India is willing to conclude such agreements with other WTO members, there appears to be no inconsistency with the most favored nation clause of TRIPs.

CSIR has begun laudable efforts to improve patent literacy amongst its scientists. This covers laws, rules, and procedures, and also an increasing awareness of the long-term benefits for the country, particularly from increased domestic R&D and productivity. These efforts are being made by ICAR also.

Although a broad discussion and debate on IPR legislation relevant to agriculture is necessary, TRIPs-compatible laws will mostly have to be in place by 2000. It is therefore time to enact the required legislation and the implementing rules and regulations, incorporating all the flexibility allowed. At the same time this exercise should be undertaken with as much transparency as possible, to allay fears raised in the public debate.

There is much to be done for IPRs in agriculture in India. The tasks that remain to be done are somewhat daunting, in light of the sharp differences of opinion amongst the stakeholders. National and international organizations have to gear up to contribute to this exercise in an urgent and meaningful way.

Appendix*

The Indian Agricultural Research System consists of 59 research institutes (including four deemed to be universities) of the Indian Council of Agricultural Research, 4 bureaux, 30 national research centers, 10 project directorates, 80 coordinated research projects with their 200 centers, and one Central Agricultural University and 28 State Agricultural Universities. This system has been providing technical inputs to government agencies on protection of intellectual properties being developed in the wake of the TRIPs Agreement, of which India is a signatory. The measures relate to legislation on plant variety protection, amendments to the existing Indian Patents Act (1970), and control over and protection of biodiversity. Until recently, there was not much appreciation of the importance of IPRs, due mainly to the concentration of research in the publicly funded institutions, and the position that the benefits of technologies generated should be available to people largely without cost. This position was held especially strongly in the agricultural research sector, whose clients were resource-poor farmers often unable to pay for new technology. The new improved technologies also had to be demonstrated before adoption. This strategy paid handsome dividends, as evidenced by the strides made by the country in accelerated food production.

With the signing of the WTO Agreement in particular, and liberalization and globalization of the Indian economy in general, a fresh and urgent look at IPR issues was undertaken. With pressing economic, trade, and technology realities emerging, especially the challenge of global competition and need for a competitive edge, a favorable environment was created for working on policy issues to conform with the TRIPs Agreement.

India, with its diverse agroclimatic zones, provides an excellent test case for developing policies on IPR which, appropriately modified, could perhaps be used by other developing countries. The identification of appropriate policies, of course, entails considerable public debate. The patenting of turmeric and recently basmati rice by industrial nations has not only caused debate in India, but has also helped in expediting action on the protection of IPRs, something long overdue.

The World Bank, through innovative and timely interventions, has been providing critical help to its member countries, especially its borrowers. The need for such help has never been greater than when the developing countries have been striving to tailor their policies to conform with the requirements of the WTO Agreement, while meeting national needs and objectives. Although the perceptions and mode of actions may vary from country to country, a broad consensus must emerge on IPR issues. There is much the World Bank can do to assist its borrowers and member countries to prepare for suitable change, or to develop new legislation in conformity with the WTO Agreement. The help could be in the form of regional/national workshops, in collaboration with NARS; human resource development; strengthening the infrastructure of IPR protection establishments such as patent offices and others; development of centers of excellence on IPRs, education of policymakers, scientists, and technologists through publication of literature; networking (electronically); access to latest developments and grants to NARS to renew and update aging infrastructure. All of these areas would have an indirect effect on the protection of intellectual property.

IPR in India

Over 65 percent of the Indian population depend on agriculture, which also contributes 27 percent of GDP and which contributes substantially to export earnings; the protection of agriculture-related IPRs is therefore an issue that will affect the lives of millions of extremely poor farmers. There has been a wide-ranging debate in India on this issue, and a consensus has emerged that the traditional and legitimate rights of farmers should be adequately protected.

---

* By S. M. Ilyas.

Plant Variety Protection

According to Article 27 of the TRIPs Agreement, a country is obliged to ensure protection of plant varieties either by patents or by an effective *sui generis* mechanism or by a combination of both. Since the Indian Patents Act (1970) excludes agriculture and horticulture from patenting, it is required to put into place a *sui generis* system by evolving legislation on plant variety protection by the year 2000.

There has been a country-wide debate involving policymakers, scientists, NGOs, and other stakeholders on the appropriate system, which suits national needs while providing for adequate protection of researchers and farmers. Although there has been a general consensus, many doubts exist over the protection of rights of farmers to use harvested materials as a seed source the succeeding season. One of the most important factors in increasing productivity has been the timely supply of inputs, especially the seeds of improved varieties. Although the public sector seed companies have been doing their best, it has not been possible to provide good seed to large numbers of farmers, particularly small, resource-poor landholders. Private seed companies are not showing much interest in supplying seeds of cereals, legumes, and oilseeds, and are instead concentrating on vegetable and fruit seeds. It is therefore necessary for farmers to use, exchange, sell, or otherwise dispose of seed produced on their farms. There is a strong case, therefore, for exempting small-scale, and resource-poor farmers. European legislators at one point considered allowing farmers with holdings below a certain size to use the harvest as a seed source. The overwhelming majority of farms in India are small, presenting a strong case for such an exception. There is a strong feeling that the legislation should not in any way hamper the momentum of indigenous research efforts to evolve new plant varieties, as that could have serious implications for India's agriculture.

The Government of India has taken due note of these concerns while proposing to introduce in Parliament a bill (Plant Varieties Protection and Farmers' Rights Act—PVP and FR) that seeks to provide adequate protection to plant breeders as well as farmers. It is understood that the draft proposals also include a Plant Varieties and Farm-

ers' Rights Protection Authority, a National Community Gene Fund, compulsory licensing provisions, and additional protection of public interest, and an Appellate Board, among others. The Government is also considering accession to UPOV 1978 version which includes breeders' and farmers' privileges. The Fourth International Conference on Plant Genetic Resources held at Leipzig (Germany) gave a global endorsement for farmers' rights. The Indian Council of Agricultural Research (ICAR) has established a National Research Centre on DNA Finger Printing at New Delhi which will be of great use in protecting materials in the event of disputes arising due to IPR. Meanwhile, pending passage of this Bill, arrangements have been made to register the new genotypes with the Bureau of National Plant Genetic Resources, New Delhi (a unit of ICAR) on a first-come-first-served basis. The registration will provide the basis for the claim of breeders' rights with the Statutory Authority under the PVP and FR Act when enacted.

Amendments in Indian Patent Act (1970)

The Indian Patent Act grants only process patents and not product patents in food, medicine, drugs, and substances provided by chemical process. Under the TRIPs Agreement countries have up to 10 years (from January 1, 1995) to bring product patent offerings up to the Agreement requirements. Notwithstanding this transition period, WTO signatory countries are required to provide a means for filing applications in the areas of pharmaceuticals and agricultural chemicals and, on fulfillment of certain conditions, the granting of exclusive rights for a period of five years or until the patent is granted or rejected, whichever is earlier. These provisions were made by amending the 1970 Act through an Ordinance promulgated in 1994. Under law, the Ordinance must be replaced by an Act of Parliament, but due to certain unavoidable reasons, it was not, so the Ordinance lapsed. According to a Government decision, the Patent Office has been receiving product patent applications in these fields, which are held unexamined until January 2005 as stipulated by the TRIPs Agreement.

The Government is actively considering introducing in Parliament a Bill amending the Indian Patents Act of 1970 to bring it in to conformity with TRIPs. The bill is expected to provide adequate safeguards to protect the interest of the people. This expectation should also allay fears that the floodgates to the patenting of genes, living organisms, and products derived from them might be opened. However, India is under no pressure to accept patenting of any life forms classified as plants and animals, although it likely has to agree to the patenting of microorganisms and products from genetically engineered microorganisms, as stipulated under TRIPs.

Presently, patent protection is available for up to 14 years, whereas it is 20 years (from the date of application filing) under TRIPs (Article 33). Under current law the onus for disproving originality of invention lies with the applicant rather than with the patent office, as in the USA. Together these result in considerable harassment of the inventor, and delay. (It takes five to seven years to obtain a patent in India.) There is a huge and growing backlog of patent applications. With the rising awareness about the need for IPR protection, greater numbers of applications will be filed by Indians as well as by foreigners, necessitating the revamping of the Patent Office. The Government is taking necessary steps in this direction.

Protection of Biodiversity

India is one of the largest reservoirs of biodiversity. An urgent need has been felt to inventory biodiversity to have a complete picture of this valuable resource. Little more than 150 of the world's 248,000 plant species are under commercial cultivation. About 20 crops provide 90 percent of the world's food, and just four of them—rice, wheat, maize, and potato—supply more than half of the daily calories. But diversity—within and between species—is the basis of most small-scale farming systems. It allows farmers to withstand vagaries of climate and disease that can severely reduce productivity. Farmers have painstakingly developed resilient and bountiful agricultural systems based on biodiversity, and their knowledge of how to work successfully in complex cultural settings. Considering agriculture as merely a matter of commodities and businesses militates against the security people derive from crop genetic diversity. Community livelihoods and the international economy can benefit from

a wider deployment of biodiversity in agriculture, including wild species and currently underutilized crops. It is reported that more than 2,000 species of native grasses, roots, fruits, and other food plants have already been classified as "Lost Crops" of Africa. Through advances in biotechnology research, resilient crops will be vital for extending cereal production into marginal lands to support growing populations.

The Wealth of India series, a monumental effort by the Council of Scientific and Industrial Research, has made serious efforts to promote nontraditional crops, but the work is by no means complete. The work is expected to take another two years. The same holds for traditional knowledge in written form and transmitted by word-of-mouth. Efforts have been made recently to assert rights over bioresources, such as the successful effort leading to withdrawal of a U.S. patent on turmeric, and earlier upholding a claim against a patent on neem oil given to a European multinational. However, the case of patenting of basmati rice has caused a public outcry and induced the Government to take steps to prevent any recurrence.

The Government is considering bringing out comprehensive legislation controlling national biodiversity to regulate access and the peoples' rights. A proactive rather than reactive strategy is required.

Role of the World Bank

The World Bank has been in the forefront of helping developing countries with burgeoning populations and dwindling resources. These countries are also caught up in the tide of globalization and are losing their competitive edge. The importance of the Bank has never been greater than now, when the developing countries are finding it increasingly difficult to cope. There is little time remaining to enact appropriate legislation to conform with the TRIPs provisions. The World Bank could assist countries in helping themselves through a process of revitalization of their institutions. Some agriculturally related areas where the role of the World Bank could be quite effective in regard to IPRs are:

1. Organizing workshops on IPR issues for policymakers, scientists, NGOs, and other stakeholders in collaboration with NARSs

2. Capacity building through short-term training and exchange visits

3. Providing networking means among the players

4. Assisting governments in strengthening statutory bodies overseeing protection of IPRs–including the modernization of patent offices by equipping them with state-of-the-art technology

5. Establishing centers of excellence that would provide assistance in the development of the latest technologies.

National Agricultural Technology Project (NATP)

The World Bank supported NATP—a $240 million project being implemented in India (the project, in fact, started in 1997 through retroactive funding). It is one of a kind, and a major milestone in the support of the World Bank to a NARS. It is expected to provide a much needed boost to the efforts of ICAR to take Indian agriculture to the next century, by increasing the capability of the scientists in frontier areas through critical support in the infrastructure, human resource building and assessment, and refinement and transfer of technology. It also seeks to provide timely help in effecting much needed organization and management reforms to increase the research management capabilities of ICAR. It is widely believed that by the time this project ends, the Indian NARS will be much stronger, better equipped, and more confident to face the challenges of the 21st century.

Notes

1. The author has worked in the Government of India, dealing inter alia with the TRIPs agreement, in the Ministry of Industry and in the Ministry of Commerce at New Delhi. The views expressed here are based on publicly available material, including newspaper reports, and are not attributable to any institution or organization with which the author is or has been associated. The author gratefully acknowledges, with the usual disclaimers, material and useful comments received from C. Niranjan Rao of Indian Council for Research on International Economic Relations (ICRIER), New Delhi. This paper has been revised incorporating comments received at a seminar held at ICRIER, New Delhi on July 9, 1998 and from the editors of this volume.

2. At the ICRIER Seminar, several participants felt that since there was no separate legislation on the subject, the parent lines of the hybrids were not legally protected in India. However, the protection provided for trade secrets or confidential information under common law and jurisprudence can be used against the unfair misappropriation of confidential information, although this would not, unlike PBRs, protect against independent discovery.

3. At the ICRIER Seminar, the representative of Monsanto categorically stated that, despite policies to encourage private sector investment in the seed sector since 1989, such investment was forthcoming only in hybrids and not in self-pollinated crops. He argued that IPR protection was required not so much to protect against theft by farmers but against misuse of other private sector seed companies.

4. It was felt in the ICRIER Seminar that such negotiating teams should have included experts in agriculture and biotechnology. Such experts were, in fact, consulted by government in formulating its position for the negotiations in WTO, although the adequacy of such consultations could be debated. Another matter raised was the need to keep the same trade negotiators to ensure continuity in negotiating strategies, a problem not unique to India alone.

5. At the ICRIER Seminar Dr. Sahai opined that the provisions of Article 27.2 of TRIPs could be used to exclude patenting of life forms. However, it was pointed out that in such a case there could be no commercial exploitation of such inventions.

6. Evidence given by Mahender Singh Tikait, president of the Vharatiya Kisan Union on September15, 1993, before the Parliamentary Standing Committee on Commerce, 1993–94, Third Report on Draft of Dunkel Proposals, Evidence, Rajya Sabha Secretariat, New Delhi, December, 1993.

7. See evidence of the officials of the Ministry of Commerce, particularly that of Shri Anwarul Hoda, now deputy director General of WTO, before the Parliamentary Standing Committee on Commerce (1993–94) as cited above.

8. See, for instance, the evidence of ICAR and IARI scientists, including Dr. M. S. Swaminathan, ex-director, ICAR, and that of Sharad Joshi, farmer activist, before the Parliamentary Standing Committee on Commerce (1993–94).

9. Evidence of such confusion was seen at the ICRIER Seminar, where many were confusing farmer's rights with the farmers' privilege. While it is yet not clear whether the farmers' rights will form part of India's legislation on plant variety protection, the latter is clearly within its ambit. With Monsanto Corporation's purchase of the terminator technology (reported on RAFI's web site, www.rafi.org), however, fears of farmers being forced to purchase expensive

seed every time are being expressed in India, and the utility of the farmers' exemption clause is being questioned.

References

Bhattacharjee, Abhijit. 1988. "New Seed Policy: Whose Interest Would It Serve?" *Economic and Political Weekly*, October 8, 2089–90.

Business Standard. "Elections Cast a Shadow on Biodiversity Law." March 5, 1998.

Buttel, F. H., M. Kenney, and J. Kloppenberg Jr. 1985. "From Green Revolution to Biorevolution: Some Observations on the Changing Technological Basis of Economic Transformation in the Third World." *Economic Development and Cultural Change* 31–55.

Economic Times. 1998. Protect Breeders' Rights Else No New Roses: MNCs. February 24.

Gupta, Anil. 1996. "Technologies, Institutions, and Incentives for Conservation of Biodiversity in Non-OECD Countries: Assessing Needs for Technical Cooperation." In *OECD Proceedings of the Cairns Conference on Investing in Biodiversity*. Paris: OECD.

Indian Express. 1997. "CSIR Goes into Documenting Drive after Turmeric Patent Triumph." August 28.

Menon, Usha. 1991. "Intellectual Property Rights and Agricultural Development." *Economic and Political Weekly* July 6–13, 1660–67.

M. S. Swaminathan Research Foundation. 1994. "Methodologies for Recognizing the Role of Informal Innovation in the Conservation and Utilization of Plant Genetic Resources: An Interdisciplinary Dialogue." Madras.

National Academy of Agricultural Sciences. 1998. "Policy Paper on Conservation, Management and Use of Agro-biodiversity." India.

Pray, Carl, and Tim Kelley. 1997. "Impact of Liberalization and Deregulation on Technology Supply by the Indian Seed Industry." World Bank, Washington, D.C.

Rao, Niranjan C. 1997. "Plant Variety Protection and Plant Biotechnology Patents: Options for India." Policy Paper 29, for UNDP-funded Project LARGE, UNDP, New Delhi.

Sahai, Suman. 1992. "Patenting of Life Forms: What It Implies." *Economic and Political Weekly*, April 25, 878–89.

———. 1994. "Government Legislation on Plant Breeders' Rights." *Economic and Political Weekly* June 25, 1573–74.

Shiva, Vandana. 1991. "Biotechnology Development and Conservation of Biodiversity." *Economic and Political Weekly* November 30, 2740–46.

———. 1993. "Farmers' Rights, Biodiversity and International Treaties." *Economic and Political Weekly* April 3, 555–60.

———. 1996. "Agricultural Biodiversity, Intellectual Property Rights, and Farmers' Rights." *Economic and Political Weekly* June 22, 1621–31.

Srinivas, Ravi K. 1994. "Power without Accountability: Draft Bill on Plant Breeders' Rights." *Economic and Political Weekly* March 26, 729–30.

Sutherland, Peter. 1994. "Seeds of Doubt: Assurance on Farmers' Privilege." *Times of India* March 15.

Watal, Jayashree. 1998. "The TRIPs Agreement and Developing Countries: Strong, Weak or Balanced Protection?" *Journal of World Intellectual Property*, 1(2, March): 281–304.

WTO. 1996. "Report of the Committee on Trade and Environment." Document no. WT/CTE/W/40, November 7 (available on the WTO web site, www.wto.org).

United States Land-Grant Colleges
Frederic H. Erbisch

The land-grant college (LGC) concept was initiated in 1862 to provide broad access to training and higher education to all, and particularly to farmers and industrial workers. The LGC has evolved into an institution that now provides a mix of vocational education and science, a compromise that addresses the social role of the university.

The LGC of today provides quality education to students from many backgrounds, it conducts research to support agriculture and industry of the future, and it provides services directly to the public. In agriculture these services include an agricultural extension service to aid the farmer on a daily basis, to provide short courses and training, and to learn directly the concerns of the farmer.

The educator in the LGC has gone from a Professor of Agriculture, Mathematics and Languages to a specialist in a field of education/research in an area that supports agriculture. The LGC has evolved from a single college/department into a number of more specialized colleges and departments within a larger university environment. The college farm is now the agricultural experiment station, where a wide range of applied and basic agricultural research is conducted. The LGC, because of its educational thrust and societal obligations, will continue to change and build for the future. One area of current change concerns intellectual properties and intellectual property rights. Recent activities may serve as a possible model for other public sector institutions.

Changes in Intellectual Property Use at Land-Grant Colleges

Because the original LGC was developed to assist the public, little attention was given to intellectual properties; generally, new intellectual properties were shared with the public with few or no restrictions. In agriculture a new crop plant variety would be made available to the farmer at no cost. Some mechanical and equipment inventions were patented and licensed, but these too were generally provided at little or no cost to the farmer/manufacturer. The farmer and industry, especially commodity groups, became accustomed to obtaining intellectual properties at little or no cost. The statement "land-grant tradition" or "land-grant way of doing business" or similar statements often refer to this "giving away" of LGC intellectual properties. These types of statements are still frequently used by LGC researchers and industrial representatives when discussing intellectual properties, with the researchers wishing to give away their ideas and industry willing to take them free of cost.

The patent system initially provided governmental protection for nonplant invention protection. The patent system protects ideas and allows the innovator to control the manufacture, use, and sale of these ideas. It was not until 1930 in the USA that patent protection was extended to include asexually produced plants only. Although more than 5 million nonplant patents have been issued in the United States, fewer than 7,000 plant patents have been granted. Plant patents have been sought mostly by private sector companies and not universities. The plant patent has had little impact on universities and their manner of handling intellectual properties.

Plant variety protection (PVP) was adopted by the U.S. in 1970. This type of protection covered sexually reproduced or reproducing plants. PVP did not protect hybrids and other nonsexually reproducing plants. The process for obtaining PVP is simple and inexpensive, largely due to the fact that this program is administered by the U.S. Department of Agriculture. The United States also became a member of UPOV, which resulted in the revision of PVP rules. Protected under present PVP rules are sexually reproducing plants and tubers. Protection includes reproduction and sale of seeds of the variety, with farmers being allowed to use or replant seed they produce. This method of protection had an impact on LGCs because it allowed them control over the distribution of new crop varieties quite inexpensively (when compared to patent protection).

In the mid 1980s, protection for certain plant materials was sought through utility patents. This type of protection has been used most in connection with the emergence of plant biotechnology. Isolation of genes, gene constructs, plasmids, and gene promoters are the basis for seeking this type of protection. Through utility patent protection, the patent owner has more control over varietal transformation than if only PVP protection were used. Obtaining patent protection is costly, time consuming, and requires experts to draft documents to obtain the best possible protection. This type of protection is used extensively by industry and LGCs.

Two events, PVP and biotechnology, have greatly increased attention to intellectual property protection and management. LGCs now routinely review their new intellectual properties to determine the best way to handle them and still serve the public. Another factor increasing use is the decline in federal government support for agricultural research and the rise of industrial sponsored research. For example, industrial sponsors of LGC research efforts usually request exclusive access to any intellectual properties developed through the use of their funds. In some instances sponsors will provide proprietary information to the researcher and demand confi-

dentiality. Even federal research support relates to intellectual property, because every government-sponsored project carries the Bayh-Dole requirements of disclosing all inventions, providing the government with a license, and diligently seeking industrial licensees for inventions. Failure to comply with these requirements can result in loss of funding, immediately and in the future.

The usual collegial exchange of materials has also been affected by these uses of intellectual property. Materials are not exchanged unless a material transfer agreement (MTA) has been executed. The MTA defines ownership of the materials being exchanged, the conditions under which they can be used, and the liability when used. The transfer of a gene construct from Michigan State University (MSU) to a company without control by an MTA was a painful lesson to a number of researchers. The gene construct was sent to a friend at a company by a professor who did the isolating. Researchers at the company determined the gene construct was valuable and patented it in the company's name. The company pays no royalties to the University, but does give the researcher a small research stipend annually and allows the researcher to use the gene construct for research purposes only. Others at the University cannot obtain the gene construct from the researcher because the company forbids this type of exchange. Any University researcher wishing to use the gene construct must contact the company and enter into an MTA. It now takes in excess of six months to obtain this gene construct from the company even though the researcher who isolated and described the gene construct is just next door. All involved have certainly become aware of intellectual properties and IPRs.

As the significance of recognizing IPRs increases, it is important to educate all those within an organization using and developing intellectual properties. Few if any professors/researchers received any training in intellectual property management in their advanced degree programs. This education is now necessary and should provide the researcher with an awareness of IPRs as they affect research activities. Additional education is generally not provided unless requested. LGCs and other academic institutions now routinely provide such educational opportunities. The Association of University Technology Managers (AUTM) was established to provide continuing education to academic intellectual property managers who, in turn, educate professors/researchers at their institutions.

Most LGCs and many other academic institutions have established technology transfer or intellectual property offices to offer this education, as well as to provide an interface between industrial and federal sponsors and academic institutions. These offices are another part of an LGC not anticipated by the founders of the system. Employed within these offices are intellectual property specialists, attorneys, accountants, and other professionals. For the LGC and the professor/researcher, these professionals handle invention disclosures, copyrights, MTAs, confidential disclosure agreements (CDA), research contracts, licenses, and the occasional intellectual property-related problem. Without this type of organization the LGC would stand to lose IPRs, as well as sometimes being in violation of others' rights.

IP Control Systems at LGCs

With the use and awareness of intellectual properties and IPRs increasing, there are numerous "situations" that arise and must be handled. One such situation reported above described the loss of compensation to the LGC and professor/researcher, and the restrictions on the professor's freedom to interact with colleagues because an MTA was not used to transfer a gene construct. Had the professor been able to consult an intellectual property specialist before sending out the gene construct, these problems could have been avoided and the company could still have received user rights to the gene construct. The intellectual property specialist would have insisted on using an MTA and licensing the gene construct with the company. If commercially successful the gene construct would have provided a royalty stream to the University and the researcher, and the University would have reserved the right for noncommercial exchange of the gene construct and patented the construct.

Appropriate policies must be in place for intellectual property specialists and others to work effectively for the LGC and its researchers. These policies define ownership of intellectual proper-

ties developed at the LGC, publication rights, and associated matters. It is through these policies that the professors/researchers and intellectual property specialists know how to operate and are able to work with those from outside the University. MSU policy on publication is to retain publication rights for faculty, researchers, and students. In one case a company approached MSU about transforming a particular type of plant. The company was willing to pay more than US$80,000 to carry out research and transformation, but no type of publication or presentation would be allowed. The University refused to accept the offer. Negotiations were conducted for over a year, during which the University held to its publication policy. The company finally conceded to the University's position.

Some groups demand ownership of any intellectual property developed under their sponsorship. MSU's policy is to assume ownership of any intellectual property developed at the university, regardless of sponsorship. The University does offer sponsors an exclusive license for any intellectual property developed under the research sponsorship. Many sponsors mistakenly believe they provide all the support for a research project, failing to recognize University support includes the research building, the research laboratory, equipment, salaries, training, and libraries. In truth the sponsor pays only a fraction of the total cost of conducting research.

The impact of intellectual properties and their protection in the biotechnology area is affecting the LGCs, their programs, and the people they serve, especially the farmers. Originally LGCs bred new crop varieties and distributed them freely to farmers. Later with PVP in place LGCs began controlling distribution of new varieties through certified seed growers, but still provided seed to farmers at low cost. LGCs are now being limited in what they can distribute, because germplasm used in their breeding programs may contain patented materials.

With the advent of biotechnology, along with industry taking a major role in biotechnology, breeding, and distribution, the traditional university breeding program may soon be gone. A major firm is providing genetically transformed soybean with the transformation units protected by patents, and which are superior to non-engineered varieties. How can LGCs compete in this market? However, if LGCs terminate traditional breeding programs, who will train the breeders of tomorrow?

Although the application of biotechnology is only beginning in the animal science area, changes in the animal industry are already affecting farmers. For example, LGC professors/ researchers working on vaccines are developing new compounds that can benefit the farmer, but because of limited markets, large companies will not license and take many of these vaccines forward commercially. LGCs do not have the funds or capability to commercialize these products, so effective new vaccines are "sitting" on the laboratory table. At MSU one professor was so frustrated with this situation he set up his own vaccine company. Although only two years old, the new company is successfully marketing vaccines large companies would not consider. Does this indicate a new trend, a trend that means LGCs may have to find ways to directly commercialize their own inventions because "big" industry is interested only in large markets? The LGC role is to serve society. How will it meet this new challenge of the large market requirement and serve society?

Lessons Learned and the Future

The major lesson to be learned is that change occurs and LGCs must adapt to these changes or lose the ability to serve society. Today's changes concerning intellectual properties affect how the LGC uses its technologies and transfers them to customers. Is this a bigger challenge than LGCs faced previously, or is it simply another step in the evolution of LGCs? The effects of this intellectual property awareness reach beyond LGCs and are being felt throughout the agribusiness world. How can LGCs and others prepare to meet this challenge?

There are four basic steps I believe need to be taken to prepare for the future:

1. Adopt appropriate policies and measures for handling and managing intellectual properties
2. Educate and prepare professors, researchers, and students
3. Develop a means to manage intellectual properties, perhaps through an office dedicated to intellectual property management

4. Provide the means for the management office to operate effectively within and outside the organization.

Many of these factors are now impacting agricultural research at the international and national levels, even in seemingly remote countries. Those entities can benefit from the ongoing experiences of the LGCs as attempts are made to accommodate change while serving traditional clients among the poor of the world.

Prior to the early 1980s there was little awareness of the importance of IPRs by universities or farmers, or the general public. Industry was well aware of the importance of protecting its intellectual property, but usually did not go beyond its own research laboratories for new technologies. During the 1980s a number of changes occurred that created an intellectual property awareness. One of these changes was the passage of the Bayh-Dole Act by the U.S. government, which gave universities the right to license technologies developed under federal funding. Another was industry reducing or eliminating its own research facilities, and needing to either have research done elsewhere or to acquire new technologies outside of the company. As well competition between companies increased as did the costs of bringing new products to market. Testing needed to meet government requirements was especially costly. Major steps in biotechnology dealing with gene research and transformation were also important in increasing intellectual property awareness.

In the United States this increased awareness led to universities licensing hundreds of inventions, resulting in a new revenue stream (royalties), the start-up of many new companies, and the employment of thousands in the new and already established licensee companies. This process of recognition, taking intellectual properties forward, and generating new companies is still evolving. As the process evolves, universities and industry are learning as they advance the transfer of technology. Those new to technology transfer and IPR will need to talk with representatives from industry and academe to learn what "works," what doesn't, and under what circumstances each approach is successful.

The key in the awareness and taking forward of intellectual properties has been the U.S. Government, which "untied" various restrictions it had imposed through its funding of research, and allowing universities and industries to work out the way to handle intellectual properties. Governments planning to support research should follow the U.S. lead in how it handles intellectual properties. The freedom to take technologies forward and allowing the marketplace to develop its standards will do much to encourage the development of IPR.

The high cost of taking intellectual properties from the research laboratory to the marketplace has made it necessary for a company to protect its properties. For example, bringing a new pharmaceutical to market will cost in excess of US$250 million. A company could not be expected to spend so much and allow others to copy the technology, or to enter into competition with it. Companies therefore closely guard their technologies.

In agriculture the advent of transgenic plants has led to the development of expensive biosafety testing. The high costs of developing transgenic varieties and biosafety testing means few companies, especially the smaller ones, can afford to undertake this work. In a few years there will be fewer than 10 companies in the world producing new transgenic varieties.

Because of the costs associated with taking a technology to the marketplace, while advocating the commercialization of new intellectual properties, managers of intellectual properties must carefully review these materials to ensure proper ownership prior to licensing. This is particularly true when using gene constructs, plasmids, and the like. An LGC cannot license to a third party any materials that belong to another party. Materials that are described in scientific articles are usually assumed to be in the public domain, and can be used by anyone without restrictions. One gene promoter was described in a scientific publication and was made available to the scientific community from a number of sources. This promoter was and is being used by a number of researchers at LGCs. However, a company had applied for a patent on this promoter and the existence of this patent application was "buried" in the United State Patent and Trademark Office for several years before being brought to the attention of the scientific community. The patent has been issued and the company owning it is allowing the use of the promoter for research purposes only. Those who have used the pro-

moter, believing it was a public domain promoter, now find this use makes their work not commercially viable. The company has said users may license these new applications of its promoter only to the company or to whom they (the company) direct. But in no case will the company allow the LGC to market elsewhere any technology containing its promoter. The researcher must know the protection status of any material developed by others and must report this to the appropriate licensing officials.

One new technology licensed by an LGC contained plasmids developed by two different universities, one in the United States and one in Australia. The inventor had informed the licensing officer who then, prior to licensing the new technology, obtained licenses from each of the universities, licenses which gave the LGC the right to include the plasmid in the LGC's technology. The two universities will each receive a share of the LGC's royalties. Another genetically modified product was brought forward which is much superior to the existing crop plants of that variety. The inventor listed plasmids from eight different sources being used to develop this new variety! Before the superior product can be licensed, permission will need to be obtained from each of the eight sources.

Licensing without this permission can result in severe penalties being brought against the LGC.

In LGCs some researchers see the approach to licensing of intellectual properties as an affront to their academic freedom. These individuals believe licensing rather than "giving away" intellectual properties will adversely affect their ability to obtain research funds and interfere with scientific publication. This is not true. The office handling intellectual properties is a service organization, and will do its best to ensure continuation of the research program and professional development of the researcher.

These researchers believe that more people will benefit if the products of their research are made freely available to everyone. This is often not true. The research results are usually not ready for the marketplace because they require more work and development. Few companies have the ability to do this R&D. Those companies that do are hesitant to spend money to take a "free" product to market. Others can copy or use the intellectual property since there is no protection, patent or otherwise.

Commodity groups that sponsor research at LGCs are requesting exclusive access to the products of the research they sponsor. They do not want the work products to be "given away." They want to use these products for the benefit of their members through limited distribution and royalty generation. In this way the commodity groups can have more funds for other research projects.

For researchers and the commodity groups a considerable amount of education is needed. This education is in basic intellectual property management and marketing. Many LGCs are doing this now through seminars, one-on-one discussions, printed materials, and other means. Probably the most important part is being able to show researchers that the system of licensing really works.

The farmer is beginning to benefit from the awareness of IPRs. Several transgenic plants are now being marketed and others will be in the near future. These new plants do or will express a "natural" resistance to harmful insects, be able to survive certain herbicide applications, tolerate low temperature, be more drought resistant, and produce greater yields. The future holds the promise of many new and exciting developments for the global community and especially the farmer. The continued recognition and exploitation of IPRs will fuel this exciting development.

5. A Model for Internationally Owned Goods

BioBanana

Gabrielle J. Persley

The World Bank, in consultation with banana-producing countries and the banana export industry, has developed a business plan for a research program that will use modern biotechnology to address the long-standing disease problems affecting this crop. The seriousness of these diseases results in the excessive use of fungicides and nematicides, with accompanying risks to human health and the environment. Recent research results suggest it should be possible to develop new varieties combining the export quality of Cavendish fruit with novel genes conferring disease resistance.

Most research and development (R&D) on banana and plantain improvement has been funded by the public sector in producing countries, including by development agencies and the CGIAR (Persley and George 1996). Because the export banana industry will also benefit from the application of newly available biotechnologies, it would be desirable to have a mechanism by which that industry could join with public sector agencies and development agencies in financing the next phase of the research, development, and delivery of new technologies. After a series of consultations during 1996–97, the World Bank invited participation in this International Banana Biotechnology Program (BioBanana). The invitations went to a wide range of institutions in banana- and plantain-producing countries, and companies involved in the banana export trade (see appendix).

The experience gained in the development of this international program may provide lessons for other groups seeking to address common agricultural problems. In particular, the feasibility study undertaken by the Bank and its partners to resolve issues relating to the management of intellectual property, anti-trust issues, and regulatory matters may provide lessons for subsequent research programs or other commodities.

The World Bank has invested about US$25 million in the banana industry in producing countries. This includes substantial investments by the International Finance Corporation (IFC), the Bank's private sector investment affiliate, related to improving the efficiency and the sustainability of the banana industry in several countries of Latin America and the Caribbean, and Southeast Asia. The World Bank has a vital interest in assisting in the control of diseases on this important crop, reducing environmental damage, and improving worker safety.

The Bank is also supporting R&D on banana and plantain through the Banana Improvement Project (BIP), which it co-sponsors and manages on behalf of the Common Fund for Commodities (CFC), and the FAO Intergovernmental Group on Bananas (FAO/IGB), as well as through its sponsorship of the Consultative Group on International Agricultural Research (CGIAR) (Persley and George 1996). Two of the CGIAR centers (International Institute of Tropical Agriculture (IITA), Nigeria, and the International Plant Genetics Resources Institute's (IPGRI) International Network for the Improvement of Banana and Plantain (INIBAP)) undertake R&D on banana and plantain, in relation to plantain breeding at IITA, and R&D networking through IPGRI/INIBAP.

The World Bank is exploring ways to work more synergistically and innovatively with the private sector in areas of mutual interest, including in biotechnology, where this can benefit the Bank's developing country member. The Bank is also examining the future needs and opportunities for developing countries to make safe and effective use of the applications of modern biotechnology to solve previously intractable problems. It is also reviewing the appropriate policies and practices on intellectual property management to enable developing countries to access and use the products and processes of modern biotechnology. Finally, the Bank is reviewing the needs and opportunities to reduce pesticide use through the wider application of integrated pest management approaches.

In initiating the Program, the World Bank proposes to

- Capitalize on the results that are emerging from R&D to date, which demonstrate that biotechnology could offer solutions to intractable disease problems
- Provide a forum for the interested parties to come together, under the aegis of the World Bank, to address common problems that are beyond the scope of any one company, country, or research institution to solve
- Demonstrate new modalities of cooperation between the public and private sectors by implementing a pilot project.

Governance and Organization

Membership of the Program would be open to any company, person, or entity engaged in the production, distribution, marketing, or controlling diseases of bananas and plantains, and any organization that is financing or intends to finance (through the Program or otherwise) activities in support of banana and plantain.

Members of the Program may include (a) public and private agencies in banana producing and exporting countries, including farmer cooperatives; (b) national and international banana producing and trading companies and other commercial companies; and (c) development agencies with interests in the social and economic development of banana and plantain producing countries.

The proposed Agreement between the Bank and the participants sets out the purpose of the Program, its objectives, research strategies, and approach, as well as the rights and obligations of the participants and the Bank, and the scale of financial contributions.

The Program will operate through a Council consisting of one representative of each member institution. The Council

1. Reviews the strategy and policies of the Program
2. Reviews the progress of the Program, including progress of individual research activities financed by the Program, and the financial situation of the Program
3. Considers and approves the annual work program and financial plan and budget for the next year
4. Advises the Bank on the appointment of members of the Scientific Advisory Panel.

The major tasks of the independent Scientific Advisory Panel are to advise on the scientific content and monitoring of research projects, as well as advising the Bank on the acquisition and disposition of intellectual property. The Panel will include persons with expertise in biotechnology, genetics, plant pathology, intellectual property management, and/or regulatory affairs, as well as in the banana industry, and the transfer of technology to developing countries. The Program's research activities are to be based on an internationally competitive research grants scheme, similar to that used in BIP.

The Program's operations, including the licensing of IPR as deemed necessary, would be financed by contributions from participants to a Trust Fund managed by the World Bank, in accordance with the provisions of the Agreement between the Bank and the Program participants. Contributions would be renewed on an annual basis for five years, subject to satisfactory progress in the research program, as measured through the meeting of agreed milestones in the research strategy and operational plan approved by the Council. The proposed allocation to the Trust Fund to achieve the R&D objectives is US$5 million per year (1998 dollars), when fully operational. This represents approximately 2.5 percent of the present annual cost of pesticide applications on bananas (currently about US$200 million).

IPR and Regulatory Issues

The Bank will require all contracting parties conducting research commissioned through the Program to disclose, transfer, and assign all technical information to the Bank. Where necessary to achieve the objectives of the Program, and after consultation with the Council and the Scientific Advisory Panel, the Bank may seek formal intellectual property protection on technical information. Participants would have the right to nonexclusive, royalty free, nonassignable licenses for use in connection with banana and plantain, on any such IPRs held by the Bank.

In order to achieve its development objectives, the Bank may also license such IPRs to nonparticipants, on terms to be determined on a case-by-case basis. In making the case-by-case determination, the Bank would consult with the Council and take into account the need to foster the Bank's objective of making the benefits of the research available in its developing member countries on reasonable terms. It would also take into account the purpose for which the technology would be used, that is subsistence agriculture or export. Any revenues accruing through the issuance of such licenses would be allocated by the Bank to the Trust Fund for the purposes of the Program. After completion of the Program, any income received from licensing would be distributed annually to former participants in proportion to their financial contributions to the Program.

The World Bank will own the intellectual property rights on new technologies, but participants will be liable for problems that may develop. It is highly unlikely that royalties will be generated during the five-year project, but may come on stream later. Licensing of proprietary technologies will be treated on a case-by-case basis. This would be the preferred option over commercial agreements being put in place from the start of a project.

The present status of any proprietary intellectual property required for banana improvement is being assessed during the preparatory stage. An intellectual property audit is also being conducted of the BIP projects managed by the World Bank to assess the current status of proprietary intellectual property relevant to bananas and emerging from the commissioned research.

The contract between the Bank and the research institutions will require the necessary regulatory approvals and registrations in the appropriate countries, for any research supported by the Program on recombinant DNA technology, and the evaluation of transgenic plants in the glasshouse and in field trials. As research moves to larger scale field trials and ultimately to commercial release, the involvement of commercial companies is anticipated. If any laboratory experimentation, glasshouse tests, or field trials of transgenic plants are to be conducted in producing countries lacking a regulatory system, to enable the safe conduct of such experimentation, the Bank would offer on request to advise on the establishment of the necessary regulatory framework.

Conclusions

Much has been made of the need for international public goods arising from international agricultural research programs. There may also be a future role for internationally owned goods, resulting from jointly funded, public/private sector programs. In the present case, an international development agency is prepared to hold intellectual property resulting from research and license it to potential users in developing countries. For the private sector the approach would be in a manner that is consistent with the agency's development objectives, and equitable to public and private sector participants in the venture. The experience of the World Bank and its partners in developing the concept of BioBanana and reducing it to practice may yield some valuable lessons for other interested parties in the international development community.

Appendix. International Banana Biotechnology Program (BioBanana)

Invitation to Participate. The World Bank invites participation in an International Banana Biotechnology Program (*BioBanana*). The purpose of the *BioBanana* program is to jointly fund research to develop biotechnology-based solutions to the major banana pests and diseases, which are, in order of priority: black Sigatoka, nematodes, *Fusarium* wilt, and viruses.

The concept of establishing a new public/private sector initiative in banana biotechnology is seen as a cost-effective way of addressing the major disease and pest problems of banana, given that there is keen interest and a critical mass of potential members ready to proceed.

Production. Banana is one of the world's top five internationally traded tropical commodities, with an annual export value of about US$5.3 billion. Total world export of banana is approximately 11.3 million tonnes from 32 countries. Export dessert bananas represent about 10 percent of total world production of banana and plantain, most of which is consumed as a staple food throughout the humid tropics of Africa, Asia and Latin America.

Disease Control. The export banana is susceptible to increasingly virulent pest and disease problems due to its narrow genetic base. Almost all the internationally traded dessert bananas are the Cavendish variety. Unlike other major commodities, there has been little success, using conventional methods, in breeding export quality products with disease resistance. The main threat to the export banana industry is black Sigatoka disease. Despite some 40 years of effort in conventional breeding, there is no commercially acceptable, resistant variety available as a replacement for Cavendish.

Progress Towards Improved Disease Control. Recent research results from the Banana Improvement Project (BIP) and other sources have demonstrated that the applications of new biotechnologies to banana enable novel genes for disease resistance to be identified, inserted, and expressed in banana. These results suggest that further rapid progress is possible to identify, introduce, and evaluate new genes for disease resistance in banana.

Results of recent R&D supported through BIP and other sources, including the CGIAR, that are important to the new initiative are:

- Commercially enabling technologies to transform banana, applicable to both Cavendish banana, the important cultivar of dessert banana, and subsistence cultivars of banana and plantain
- Two novel genes potentially conferring resistance to banana bunchy top virus and bract mosaic virus

- Transgenic Cavendish plants containing the novel genes for potential virus resistance. These plants are presently being evaluated in the greenhouse and will undergo field evaluation at several sites worldwide in 1998–99
- Sources of resistance to black Sigatoka and to nematodes have been identified in banana and plantain breeding programs at the International Institute of Tropical Agriculture (IITA) in Nigeria, and at breeding programs in Brazil, Guadeloupe, and Honduras.

The first priority is to build on these results to introduce into Cavendish banana cultivars resistance to the major fungal disease black Sigatoka, which affects both dessert bananas important to international trade, and cooking bananas and plantain, important as subsistence food particularly in Africa and Latin America.

Governing Council. The *BioBanana* Program would be guided by a Council consisting of one representative of each member institution. The Council would play a key role in determining the policy and strategy of the Program. The World Bank envisages that participants will represent a cross section of industry and public sector institutions involved in banana and plantain production, trading, research, and development activities.

World Bank Trust Fund. The Program's operations would be financed by annual contributions from its members to a Trust Fund.

Scientific Advisory Panel. The Panel would advise the Council and the World Bank on the technical content of the research program. Projects would be selected on the basis of internationally competitive grants.

The aim of the Program is to mobilize the best available expertise worldwide to tackle the threats posed by diseases to banana and plantain production.

Reference

Persley, G. J., and P. George, ed. 1996. *Banana Improvement: Research Challenges and Opportunities.* Environmentally Sustainable Development, Agricultural Research and Extension Group Series: *Banana Improvement Project Report 1.* Washington, D.C.: World Bank.

6. Summary and Implications for the World Bank

Uma Lele, William Lesser, and Gesa Horstkotte-Wesseler

Private investment in agricultural biotechnology research by seed companies is increasing rapidly. The privatization of intellectual property, and the associated emergence of the private sector as the major force in agricultural technology generation, is beginning to have a profound impact on farmers and researchers in industrial and developing countries. There is a qualitative difference, however, in the impact of biotechnology and intellectual property rights (IPRs) on the farmers of developing countries. Nearly 100 million poor households in developing countries depend on agriculture for their livelihood. Most of them are resource-poor farmers. From their perspective it is not simply the productivity growth in agriculture that is important; agriculture provides food security and their livelihood. These farmers conserve their own seed as planting material, thus ensuring food security. Technological changes have helped to conserve ever-decreasing land and water resources, and to mobilize new resources into agriculture, particularly for the benefit of the resource-poor farmers. New "designer" genes help make this possible. They contain built-in insect and pest resistance. They enable cultivation of crops on previously underutilized and degraded lands. The genetic characteristics of the traditional plant material used by resource-poor farmers can contribute to the development of new transgenic materials. This makes the poor important guardians of biodiversity even in the new context of a modernizing agriculture. It is this intersection of indigenous knowledge and resources, the vulnerability of the food-insecure households needing access to vital planting material, and modern science, that makes IPRs in agriculture especially important, relative to IPRs in industry, including the pharmaceutical sector. Resource-poor farmers must have free and easy access to new agricultural biotechnologies, to ensure food security and improved rural living standards.

Public sector research has traditionally been the source of new technologies for small farmers. The development of new biotechnologies, however, requires massive investments. Cost recovery through proprietary science constitutes an important incentive for private investments in research. With rapidly growing private investment in agricultural research, important questions emerge:

- How will access to new technologies be assured to resource-poor farmers, free or at low cost?
- What role will public sector research play in future agricultural development in developing countries?
- What types of public-private partnerships will evolve to achieve social objectives?
- What incentives are needed to ensure continued public and private investment in agricultural research?

These questions have acquired a special urgency as public sector investment in agricultural research has decreased in recent years, at the national and international levels.

The World Bank has a major stake in the answers to these questions, being the largest international funding agency of public sector agricultural research in non-OECD countries. The Bank has a major challenge to assist its client countries, given (a) the wide range of country sizes, (b) the varied stages of NARS development, and (c) the highly dynamic state of technical change and varied institutional arrangements. It is a difficult and complex task to establish IPR regimes in developing countries, because of the diversity of stakeholders, each with differing interests, and the diversity of international agreements related to IPR. The Convention on Biodiversity, WTO, and the newly emerging farmers' rights regarding plant genetic material, are not easy to reconcile. Notwithstanding these complexities, the fact remains that developing countries that are signatories must meet WTO requirements by 2000 and in some cases 2005.

The World Bank proposed a workshop as a means to explore the implications of these developments for the Bank's advice and lending policy to its member countries. It sought to bring together informed people from industrial and developing countries, public institutions, the private sector, WTO, WIPO, and the CGIAR centers to

- Explore recent developments in IPRs as applied to agriculture
- Explore implications for the World Bank's future research and operational roles in this important area, including the design and implementation of the Bank-supported investments in the NARS of its client countries.

Participants were selected on the basis of their knowledge of IPR and agricultural research issues, representative of a broad spectrum of opinion. The private sector representatives strongly supported the establishment of broad IPR systems. Representatives from India outlined the complexity of the challenge in developing a national IPR policy and institutions, given the competing demands of the different stakeholders. Representatives of Brazil outlined the challenges of the public agricultural research system in the face of growing acquisition of seed companies by the private sector. The CGIAR representatives highlighted the range of issues affecting the Centers, which operate at the crossroads of industrial and developing countries, small farmers, national and international programs, and the private sector.

Given the highly dynamic nature of technical change and the growing role of the private sector particularly in biotechnology, developing country scientists in public sector research systems must have continued access to new knowledge to perform effectively their public good function. The effects of the new environment for continued access of developing country scientists and farmers to new technologies, and their implications for the World Bank assistance to its clients in technology development and access, are discussed below.

Agricultural Research Environment

Until quite recently, food crop research was almost entirely supported by the public sector, even in industrial countries. Commodities such as coffee and sugar have long been the preserve of industry groups, with public support of food crop research typically being between 0.5 and 2 percent of GDP on a national basis. The national research investments, with approximately US$400 million invested annually in the CGIAR system, supported a broad-based program able to meet many of the global food production technology needs. There are many reasons for the large public investment, including the clear public benefit of safe and inexpensive food. Farms tended to be too small, even in industrial countries, to support their own research, so the public sector became the investor of last resort. Public sector interest in agriculture has also tended to be a product of the political power of the large farm bloc that benefited from such research. The private sector had little incentive to invest, because with open-pollinated crops dominating most food species, appropriation of benefits was nearly impossible.

The private sector became interested in agricultural research when corn hybrid technology was developed in the 1930s. Within a few decades, wherever private involvement was permitted, the hybrid corn sector was almost universally privatized in major producing areas. Yet that was the exception. Hybridization provided a means of appropriation of value, while increasing the yield sufficiently to justify a premium price over open-pollinated varieties. The private sector has long dominated the agricultural machinery and chemicals sectors. Plant breeding was an exception for many years.

That situation changed in the 1970s with the expansion of Plant Breeders' Rights (PBR), a patent-like system applied to plants. That legislation is associated with the creation of about 100 seed-breeding companies in the United States, and increasing private investment in seed breeding (Butler and Marion 1985, table A-2). Adoption of PBR in developing countries lagged, and by 1994 only two countries (in Latin America) belonged to UPOV, the international convention for PBR, with only Argentina showing any investment effects of PBR, largely because of enforcement (Jaffe and van Wijk 1995). Many countries and regions, such as India and southern Africa, limited private involvement in the seed sector until quite recently.

India, for example, prohibited the importation of seed for a wide range of crops until the New Policy for Seed Development was adopted in 1988. By 1993 public and private investment in corn breeding, as measured by numbers of researchers, had grown to almost even, 108.5 vs. 92 (Pal, Singh, and Morris 1998, table 14.3). India continues to lack PBR legislation that restricts private sector investment in open-pollinated seed breeding. Zimbabwe followed a different legislative path, in part due to the influence of the large size of farms, and mostly white estate farmers. Zimbabwe adopted a national PBR law in 1973, and by 1995, five multinational firms each invested more in corn breeding than the government and CIMMYT combined (Rusike 1998).

The maize seed industry in Malawi progressed along entirely different lines, largely because of the dominant smallholder sector. Small producers preferred open-pollinated varieties that could be used as a seed source, and white flint varieties that were resistant to insect damage under local storage conditions. The hybrid sector exported yellow dent varieties worldwide, leaving relatively little material available for use in the country. Private investment began in 1989 with a single firm (Cargill), until the arrival of Pannar in 1993 following a significant deregulation of access and profit repatriation. The public sector, using materials from CIMMYT, remains the major producer of varieties (Rusike and Smale 1998).

Private sector involvement with food crops did not accelerate until the 1990s, however, when three forces coincided. The most significant was the development of agricultural biotechnology, with the first commercial release occurring in China in the early 1990s. By 1998, an estimated 27.8 million hectares (excluding China) of transgenic crops were planted worldwide, with herbicide and insect resistance the leading traits. That same year transgenics constituted 36 and 22 percent of the U.S. soybean and corn acreage, respectively. Argentine soybean areas were 55 percent transgenic, as was 45 percent of Canadian canola area (James 1998). Outside of China, essentially all of the materials involved are privately owned (CGIAR System Review Secretariat 1998, 87).

Also in the 1990s, the TRIPs requirements of the World Trade Organization were adopted, which established certain minimal levels of IPR protection that countries must have in place by the end of 1999 (Lesser and others, this volume). That mandate led to the adoption of PBR by a number of countries, so that membership in UPOV, the international PBR convention, grew to 37 in 1998 (from 19 in the early 1990s). Several other countries, including Brazil, have made plans to join in the near future (Sampaio, this volume). The patent protection of plants, an allowed alternative to PBR under TRIPs, has, however, remained available only in a limited number of countries.

The WTO agreement was but one aspect of a general opening of economies and consequent globalization, in the wake of the advance of capitalism following the collapse of communism as a major economic force in the 1980s. One prominent example among major agricultural nations is Brazil, which went through a significant economic restructuring in 1993, including the adoption of a new IPR law in 1996, and PBR legislation in 1997, along with a major reduction in foreign investment restrictions. Within a short period, Monsanto had purchased 16 national seed companies, and has entered protracted negotiations with EMBRAPA, the agricultural ministry, regarding the introduction of transgenic crops using EMBRAPA's proprietary varieties.

This ongoing transformation of food crop research from public to private dominance is raising a number of fundamental issues for national programs, for the CGIAR, and for the World Bank. Perhaps the most fundamental of these is the access of resource-poor producers to the new biotechnologies. There are a number of reasons to question the extent to which the market would be able to supply that group, without several types of intervention. Issues include the general size economies of new technologies, the additional management requirements of the new technology, greater cash flow aspects implying greater risk for resource-poor producers, the provision of stewardship for many products (such as resistance-delaying approaches for *Bacillus thuringiensis* crops), the availability of varieties meeting small farmer needs, and IPR issues in general. As a group these matters are imposing and can better be approached in their constituent components, with emphasis on IPR and public/private sector partnerships.

The CGIAR, after a lengthy period of abstracting itself from the matter, is actively considering the management of IPRs. Essentially, the past practice of unconditional release of CGIAR inventions is now recognized as less than satisfactory for a number of reasons, including the realization that it denies the Centers any control over the inventions. Consequently, a decision was made at the Mid-Term Meetings in Brasilia in May 1998 to reevaluate that policy, and to establish a centralized IPR entity with as yet undefined responsibilities (for further details see Ninnes, this volume). The matter was revisited in the Third System Review (CGIAR System Review Secretariat 1998), and incorporated into Recommendation 4, which reads in part regarding IPRs and public/private sector partnerships):

The Panel recommends an integrated gene management approach based on

- A central coordinating and servicing unit for advising IARCs and appropriate NARS
- Patenting processes and new varieties, and entrusting their use under free licensing
- A legal entity that could hold CGIAR patents.

In related actions, the Systems Review panel recommended the strengthening of CGIAR links with the international private sector, but noting that doing so would change the CGIAR's nonpartisan, nonideological character. They proposed the creation of a foundation to manage funds from nontraditional sources such as royalties from IPRs, and grants for research not compromising the public interest. The same foundation could potentially hold title to CGIAR IPRs.

As these matters advance a number of additional detailed issues will likely require further discussion, including:

- The benefits of a single service unit in Washington compared with individualized support at the Center level.
- Potential liability associated with IPR ownership.
- Costs and funding sources for IPR protection. US University patenting and licensing centers have typically found that a 10-year period is required before the process breaks even and begins to generate funds.
- Funding sources to defend IPRs from infringement.

World Bank client countries can learn from the CGIAR experience. Although that is necessary, however, it is not sufficient. First, not all the Bank's client countries are members of the CGIAR. Second, the bilateral challenges the Bank's client countries face relative to the private sector, as well as the CGIAR, are becoming increasingly complex. Finally, the client countries are directly accountable to their national stakeholders whose interests they must protect. The CGIAR, although with a mission of international public goods, is largely funded by industrial countries that do not often agree on key issues of interest to the CGIAR, either amongst themselves or with developing countries. The Bank's client countries must therefore have an independent capacity to address the challenges, as they perceive them.

The World Bank's past project investments were largely directed to developing public sector research organizations and institutions, and to training. Because the private sector has become a major supporter of agricultural biotechnology and related research, the Bank now has a major stake in ensuring that the rules of the game for public/private sector partnerships are structured in developing countries in such a way that they can ensure fulfillment of the Bank's mission of sustainable poverty alleviation. This means the small, resource-poor farmers must have continued access to the needed materials and knowledge as science and technology advance. Indeed, the public sector is in a partial crisis worldwide as it is attempting to determine its new role in the increasingly privatized agricultural research arena. Within the context of declining public sector research funding, private sector control, based on IPRs and material transfer agreements and, where permitted, through partial private sector priority-setting through its research funding support, have left public sector researchers uncertain of their appropriate role. The establishment of partnerships with clear public/private roles would help clarify the matter.

One area where these issues are evident is the public sector seed industry. The entry of the private sector into corn breeding in several developing countries was discussed earlier. With PBR being extended to most countries, the expectation now is for greater private sector involvement in open-pollinated varieties in many countries. This is an area of importance for the Bank funding of seed production. Overarching issues include the appropriate incorporation of the private sector with a quintessentially public good product, such as open-pollinated seeds, where the transaction costs of exclusion can be quite high. This is particularly true of the small farm sector where saved seed (both direct and through local market sales) is a major seed source. By contrast monopoly public sector control of the sector had, in many cases, led to a slow release of improved varieties and often inadequate and untimely distribution. Reform often required legislative changes, plus the abandonment of subsidies that tended to entrench the public monopoly.

The structure of agricultural research is evolving rapidly to one with a growing private sector component. In agricultural biotechnology the private sector is the driving force and is establishing the priorities. Public sector institutions wishing to work in this area must identify means of developing partnerships with private firms for access to expertise, materials, and funding. Thus a high priority of the World Bank should be to ensure the fair structuring of those partnerships for its global poverty alleviation mission. The Bank has currently little in-house IPR capacity to assist its task managers responsible for the design and supervision of the World Bank-financed projects. The Bank is now developing greater IPR capacity as an outgrowth of the demand from client countries such as Brazil and India. IPR is a component, but only one component, of those Bank-borrower partnerships. Relevant for national research and the Bank, but largely outside the interests of private firms, are the issues of development and transfer of materials appropriate for resource-poor farmers, in the context of the growing importance of biotechnology.

The private sector is having different effects on more traditional parts of agricultural research such as seed breeding. For example, private companies have entered into partnerships on a limited basis, particularly where hybrid seed production is involved, and where benefits can be appropriated. The rapid spread of PBR likely foreshadows a broader entry of private firms into open-pollinated seed breeding. Ongoing reform projects of national seed programs have been developed to accommodate private sector concerns, but private firms will not satisfy the needs of all farm types. This means that partnerships and split responsibilities require further formulation in this area as well.

One of the early tasks for increased Bank involvement in the complex of issues of public/private sector partnerships and IPR will be identification of expertise in these areas. Various possibilities exist, from additional in-house staff (uncertain given the present fiscal climate), to increased use of consultants, to joining with the proposed CGIAR central resource, although that will take time to develop. The latter possibility is insufficient because the Bank deals with client countries whose challenges in dealing with IPRs are different from those of the CGIAR.

Possible Future Role of the World Bank

The World Bank can assist its clients in various ways. Several conclusions emerging from the discussions at the workshop are outlined below.

Gaps in Knowledge and Need for Further Research

The World Bank could become a catalyst for promoting policy research in IPR issues of high priority in agriculture, and in assessing their impact (for example, on investments, prices, supply of technologies, distributive effects, human and scientific capacity development) among industrial countries, developing countries, and the CGIAR centers. It could support and refine operationally an international index (such as the one developed by Ginarte and Park (1997), or the one by Sherwood (1997)) of intellectual property regimes and improvements/deterioration in them on an annual basis. This would serve as a means to measure progress and to identify alternative solutions, in much the same way that the Bank uses macroeconomic and other indicators of interest to investors. The World Bank could also monitor the increase in private investments in agricultural research and technology transfer, and clarify its causes as it relates to IPRs.

Approaches for Strengthening National IPR Systems

A preliminary list of areas in which bilateral and multilateral organizations might assist developing countries in reforming their IPR regimes was suggested in an electronic conference on IPRs and Economic Development organized by the World Bank in May 1998. In view of the CGIAR decision at the MTM in Brasilia in May 1998 to establish a system-wide legal advisory facility, two efforts could combine forces as appropriate depending on the resources the World Bank might able to devote to this effort to:
- Identify and understand the IPR related issues the Bank's borrowers face
- Develop strategies for assisting in the reforms of IPR laws and enforcement procedures, taking into account the "TRIPs standards of protection" (for the CGIAR, identify the implications of the state of affairs in developing countries for CGIAR operations)
- Promote cost-saving administration of patents and trademarks
- Educate policymakers and the general public about the complex trade-offs surrounding IPRs, and what TRIPs-related IPR reforms will and will not accomplish
- Sensitize researchers and small- and medium-sized firms in developing countries on emerging opportunities, and promote linkages between research institutions and private entrepreneurs
- Review IPR legal systems for their effectiveness by making available judicial reform specialists
- Evaluate the operations of regional IPR offices to identify best practices
- Finance placing of mid-career national science managers of developing countries in experienced patent offices of industrial countries (for

example, as in the Agricultural Technology Project in Brazil) as a way of enhancing capacity of the national systems.

Identifying and Removing Practical Hurdles to Implementing Effective IPR Systems

The World Bank could:
- Diagnose needs, facilitate consensus among important stakeholder groups in each country, and help develop action plans
- Develop a network of international experts and promote regular consultations through an electronic forum to support the work of its task managers and client countries–see list of participants at the end of this volume
- Develop standard contracts for licensing component materials
- Operationalize these contracts on a pilot basis through ongoing World Bank projects to stimulate private investment in agricultural research in developing countries.

Incorporation of IPRs within the Scientific and Related Communities

The World Bank could:
- Create sufficient in-house capacity on a long-term basis through regular staffing to perform a catalytic role of mobilizing expert consultants from outside to assist its task managers in developing investment projects in developing countries, for example, through the training of those concerned with IPRs (scientists, policymakers, administrators, patent examiners, and judges)
- Assist developing countries in acquiring negotiating skills for technology transfer and material transfer agreements (needed in the CGIAR centers as well as the NARSs) through enhanced information flow.

World Bank Training Needed

Training emerged as an important area for World Bank involvement in IPR. The Bank often includes a training component in the investment projects it finances in developing countries, but for IPR there is a serious need for exposure and training of Bank staff. Bank task managers need to understand the implications of IPRs, as well as to participate in a network of outside specialists from whom they can seek help to ensure they are able to impart sound advice to borrowing countries. Bank staff need to know more about institutional arrangements (for example, in U.S. land-grant colleges and USDA as well as in other OECD countries) so they can apply some of this experience when working with developing countries. Bank managers dealing with agriculture need to have a deeper knowledge of the significance of IPRs. This would help ensure quality of public goods research investments designed by task managers, and funded by the World Bank.

The Bank needs to have a core group in its Rural Development Family who, while not experts themselves, would remain at the cutting edge of IPR developments. They would work with task managers in all the regions to develop appropriate responses through policy advice and project design. This means exposure to the more complex problems of countries such as Brazil and India, where the Bank is currently lending. These countries have large and capable national systems, and yet have had to take the lead in working through IPR problems because they had urgent demands arising out of the rapidly changing international circumstances. They face highly complex social issues–partly as a result of having large populations of poor rural households, whose "public goods" rights they must protect. But the Bank must also work with smaller and less developed countries. The latter have less internal capacity to keep up with the challenges in a rapidly moving field, and less capacity to articulate their needs to the World Bank. The challenges in providing assistance are quite different in middle-income, larger countries with an already well-established critical mass of internal capacity, as distinct from the smaller, poorer countries.

Training, therefore, has to have a two-tiered approach: First there must be an urgent effort to develop the Bank's small-core capacity at the center on agricultural IPRs with a long term view, honing the skills of Bank task managers to allow them to do a better job in this increasingly important and complex area. And second, development of appropriate responses in institutional development for public/private partnerships requires greater attention at the stages of design and implementation of Bank-financed investments. This calls for an appraisal of policy and institutional arrangements in borrowing countries and changes needed therein, in addition to the provision of funds for training of key individuals from developing countries on a priority basis to enhance national efforts.

References

Butler, B. J., and B. W. Marion. 1985. "The Impacts of Patent Protection on the U.S. Seed Industry and Public Plant Breeding." U. Wisconsin, NC-117, Monograph 16.

CGIAR System Review Secretariat. 1998. "The International Research Partnership for Food Security and Sustainable Agriculture." Third System Review. Washington, D.C., Oct. 8 (see http://cgreview.worldbank.org/cgrevrep.htm).

Ginarte, J. C., and W. G. Park. 1997. "Determinants of Patent Rights: A Cross-National Study." *Research Policy* 26:283–310.

Jaffe, W., and J. van Wijk. 1995. "The Impact of Plant Breeders' Rights in Developing Countries." Inter-American Institute for Cooperation on Agriculture, Univ. Amsterdam.

Pal, S., R. P. Singh, and M. L. Morris. 1998. "India." In M. L. Morris, ed., *Maize Seed Companies in Developing Countries*. Boulder, Col.: Lynne Rienner Publishers.

Rusike, J. 1998. "Zimbabwe." In Morris, ed., *Maize Seed Companies in Developing Countries*.

Rusike, J., and L. Smale. 1998. "Malawi." In M. L. Morris, ed., *Maize Seed Companies in Developing Countries*.

Sherwood, R. M. 1997. "Intellectual Property in the Western Hemisphere." *Inter-American Law Review* 28:566–95.

Appendix. Workshop Participants and the Creation of an IPR Network

At the mid-1998 workshop on intellectual property rights in agriculture, many of the participants felt that the creation of an informal (mostly electronic) network would:
- Assist developing countries in getting ready to meet the WTO/TRIPs requirements by the year 2000
- Help develop their capacity to stimulate further investments in agricultural research through the intellectual property environment that such a network would provide.

What Can or Should Such a Network Do?

- Network members can make valuable information available to others (for example, interesting papers; new experiences).
- Developing country members can obtain advice on IPR-related issues from other network members.
- Institutions can disseminate information on their activities (for example, on training: MSU, WIPO).
- Network members can provide comments on papers (for example, papers coming from the CGIAR centers).
- If network members encounter specific (practical?) problems related to IPRs, they can obtain advice from other network members.
- Network members can point out links to other sources that might have more relevant information (for example, TechNet; other individuals not yet included).
- Network members can debate certain topics (e-mail conference on well-defined topics).

- Focus on the role of the World Bank as a facilitator.

What Can or Should Such a Network Not Do?

- Debate issues in general
- Provide legal advice
- Provide business opportunities
- Duplicate efforts (for example, developments in international forums can best be accessed directly through these forums).

Contact Addresses of Workshop Participants

Ameur, Charles
World Bank
1818 H Street NW
Washington, DC 20433, USA
Tel.: (1-202) 473 2349;
E-mail: cameur@worldbank.org

Anderson, Jock
World Bank
1818 H Street NW
Washington, DC 20433, USA
Tel.: (1-202) 473 0437
E-mail: janderson@worldbank.org

Barton, John H.
Stanford University, Law School
Crown Quad 237
Stanford, CA 94305-8610
Tel.: (1-650) 723 2691
E-mail: jbarton@leland.stanford.edu

Bennett, A. Rick
USDA, ARS
Rm 102, Building 005 Barc-West
10300 Baltimore Blvd.,
Beltsville, MD 20705, USA
Tel: (1-301) 504 5705
Fax: (1-301) 504 5298
E-mail: rick.bennett@usda.gov

Berthaud, Julien
ORSTOM
Ave Agropolis, BP 5045
34032 Montpellier, France
Tel.: (33-4) 67 41 61 65
Fax: (33-1) 40 36 23 85
E-mail: julien.berthaud@mpl.orstom.fr
Fax: (1-301) 504 5060
E-mail: djb@ars.usda.gov

Braga, Carlos Primo
World Bank
1818 H Street NW
Washington, DC 20433, USA
Tel.: (1-202) 473 3927
E-mail: cbraga@worldbank.org

Bragdon, Susan
IPGRI
Via delle Sette Chiese 142
00145 Rome, Italy
Tel: (39-6) 51 89 24 00
Fax: (39-6) 575 0309
Email: s.bragdon@cgnet.com

Brandao, Guilherme Euclides
CNPq
Av. W3 Norte Quadra
509-Bloco A-Sala 422
Brazilia, D F., Brazil
Tel.: (55-61) 348 9386
gbrandao@cnpq.br

Byerlee, Derek
World Bank
1818 H Street NW
Washington, DC 20433, USA
Tel.: (1-202) 458 7287
E-mail:dbyerlee@worldbank.org

Carvalho, Nuno
WTO
Geneva, Switzerland
Tel.: (41-22) 739 5111
E-mail: nuno.carvalho@wto.org

Cohen, Joel
ISNAR
P. O. Box 93375
2509 AJ The Hague, The Netherlands
Tel: (31-70) 349 6158
Fax: (31-70) 381 9677
E-mail: j.cohen@cgnet.com

De Haan, Cornelis
World Bank
1818 H Street NW
Washington, DC 20433, USA
Tel.: (1-202) 473 0347
E-mail: cdehaan@worldbank.org

Dodds, John
ICARDA
P.O. Box 5466
Aleppo, Syria
Telex: (492) 331208, 331263, or 331206
ICARDA SY
Tel.: (963-21) 213477, 225112, or
225012
Fax: (963-21) 213490, 225105, or
744622
E-Mail: j.dodds@cgnet.com

Dryden, Jr. R.N. (Sam)
Big Stone Partners
1634 Walnut Street, Suite 301
Boulder, CO 80302-5400, USA
Tel.: (1-303) 449 9696
Fax: (1-303) 449 9699
E-mail: bigston1@ix.netcom.com

Echeverria, Ruben
Inter-American Development Bank
Environment Division, Sustainable
Development Department,
1300 New York Ave., NW
Washington, DC 20577, USA
Tel: (1-202) 623 1888
Fax: (1-202) 623 1786
E-mail: rubene@iadb.org

Erbisch, Frederic H.
Michigan State University
Office of Intellectual Property
238 Hannah Admin. Bldg.
E. Lansing, MI 48824, USA
Tel.: (1-517) 355 2186
Fax: (1-517) 432 3880
E-mail: erbisch@pilot.msu.edu

Espinosa, Octavio
Development Cooperation (Intellectual
Property Law) Department
34, Ch. Colombettes
1211 Geneva 20, Switzerland
Tel.: (41-22) 338 9281 (direct)
Fax: (41-22) 338 9898
E-mail: octavio.espinosa@wipo.int

Fink, Carsten
World Bank
1818 H Street NW
Washington, DC 20433, USA
Tel.: (1-202) 458 7648
E-mail: cfink@worldbank.org

Forno, Douglas
World Bank
1818 H Street NW
Washington, DC 20433, USA
Tel.: (1-202) 473 9406
E-mail: dforno@worldbank.org

Horstkotte-Wesseler, Gesa
World Bank
1818 H Street NW
Washington, DC 20433, USA
Tel.: (1-202) 458 7915
E-mail: ghorstkottewesse@worldbank.org

Ilyas, S.M.
ICAR
Krishi Bhavan
New Delhi 110 001, India
Tel: (91-11) 338 9526
Fax: (91-11) 338 9526/7293
E-mail: smilyas@icar.delhi.nic.in

Ives, Catherine L
Michigan State University
Agricultural Biotechnology for Sustainable
Productivity (ABSP) Project
324 Agriculture Hall
E. Lansing, MI 48824, USA
Tel.: (1-517) 432 1641
Fax: (1-517) 353 1998
E-mail: ivesc@pilot.msu.edu

Lee, Karen
WIPO
Geneva, Switzerland
Tel: (41-22) 338-9960 (direct)
Fax: (41-22) 338 9770
E-mail: karen.lee@wipo.int

Lele, Uma
World Bank
1818 H Street NW
Washington, DC 20433, USA
Tel.: (1-202) 473 0619
E-mail: ulele@worldbank.org

Leskien, Dan
P.O. Box 500952
D 22709 Hamburg, Germany
Tel: (49-40) 431 89766
Fax: (49-40) 431 89709
E-mail: Dan_Leskien@compuserve.com;
100703.771@compuserve.com

Lesser, William Henri
Cornell University
405 Warren Hall
Ithaca, NY 14853, USA
Tel.: (1-607) 255 4595
Fax: (1-607) 255 6696
E-mail: whl1@cornell.edu

Lewis, Josette
USAID
G/EGAD/AFS
Washington, DC 20523-2110, USA
Tel.: (1-202) 712 5592
Fax: (1-202) 216 3010
E-mail: jlewis@usaid.gov

McMahon, Matthew
World Bank
1818 H Street NW
Washington, DC 20433, USA
Tel.: (1-202) 473 8586
E-mail: mmcmahon@worldbank.org

Ninnes, Peter
CIMMYT
Apdo. Postal 6-641
06600 Mexico, D.F., Mexico
Tel.: (52-5) 726 7503
Fax: (52-5) 726 7585
E-mail: pninnes@cimmyt.mx

Ozgediz, Selcuk
World Bank
1818 H Street NW
Washington, DC 20433, USA
Tel.: (1-202) 473 8937
E-mail: sozgediz@worldbank.org

Persley, Gabrielle J.
Biotechnology Alliance Australia
P.O. Box 1101
Toowong, Brisbane, Australia
Tel: (61-7) 3365 4939
Fax: (61-7) 3365-7093
E-mail: g.persley@mailbox.uq.edu.au

Reifschneider, Francisco
EMBRAPA, SAIN
Parque Rural, W/3 Norte-Final
Brasilia-DF 70770-901, Brazil
Tel.: (55-61) 274 5000
Fax: (55-61) 272 4656
E-mail: sci@sede.embrapa.br

Salerno, Rosina
World Bank
1818 H Street NW
Washington, DC 20433, USA
Tel.: (1-202) 473 8914
E-mail: rsalerno@worldbank.org

Sampaio, Maria Jose Amstalden
EMBRAPA, SAIN
Parque Rural, W/3 Norte-Final
Brasilia-DF 70770-901, Brazil
Tel.: (55-61) 273 6454
Fax: (55-61) 347 1041
E-mail: sampaio@sede.embrapa.br

Serageldin, Ismail
World Bank
1818 H Street NW
Washington, DC 20433, USA
Tel.: (1-202) 473 4502
E-mail: iserageldin@worldbank.org

Seth, Ashok Kumar
World Bank
1818 H Street NW
Washington, DC 20433, USA
Tel.: (1-202) 458 1438
E-mail: aseth2@worldbank.org

Shear, Rick
Monsanto
800 North Lindbergh Blvd.
St Louis, MO 63167, USA
Tel.: (1-314) 694 3215
Fax: (1-314) 694 9009
E-mail: richard.h.shear@monsanto.com

Sherwood, Robert M.
International Business Counsellor
7617 Leith Place
Alexandria, VA 22307-1928, USA
Tel.: (1-703) 768 4118
Fax: (1-703) 768 4630
E-mail: rmsherwood@ibm.net

Simon, Elke
Hoechst Schering AgrEvo GmbH
Hoechst Works, K 801
D-65926 Frankfurt am Main, Germany
Tel: (49-69) 305 6055
Fax: (49-69) 305 2200

Slingenberg, Max
Agricultural Attache
The Netherlands Embassy
4200 Linnean Ave. NW
Washington, DC 20008, USA
Tel.: (1-202) 274-2718
Fax: (1-202) 244 3325
E-mail: slingenberg@was.minbuza.nl

Terry, Eugene
World Bank
1818 H Street NW
Washington, DC 20433, USA
Tel.: (1-202) 473 8544
E-mail: eterry@worldbank.org

Traxler, Greg
Auburn University
203 Comer Hall
Auburn University, AL 36849, USA
Tel.: (1-334) 844 5619
Fax: (1-334) 844 5639
E-mail: gtraxler@acesag.auburn.edu

van Schoonhoven, Aart
CIAT
Cali, Colombia
Tel.: (57-2) 445 0000
Fax: (57-2) 445 0073
E-mail: a.schoonhoven@cgnet.com

von der Osten, Alexander
CGIAR
World Bank
1818 H Street NW
Washington, DC 20433, USA
Tel.: (1-202) 473 8918
E-mail: avonderosten@worldbank.org

Weiskopf, Beate
GTZ
Dag- Hammarskjöld-Weg 1-5
Postfach 51 80
65726 Eschborn, Germany
Tel.: (49-6196) 79 1432
Fax: (49-6196) 797173
E-mail: beate.weiskopf@gtz.de

Winkelmann, Donald L.
TAC
355 East Palace Avenue
Santa Fe, New Mexico 87501, USA
Tel.: (1-505) 988 1284/1285
Fax: (1-505) 988 1285
E-mail: tacwink@newmexico.com; tac-chair@cgnet.com